TRANSDERMAL
MAGNESIUM
THERAPY

TRANSDERMAL MAGNESIUM THERAPY
A new modality for the maintenance of health.
by Mark Sircus, Ac., OMD

First Edition published in 2007

The author of this book does not dispense medical advice or prescribe the use of any technique as a form of treatment for physical or medical problems without the advice, either directly or indirectly, of the reader's own licensed counsel on such matters. The intent of the author is only to offer information to help each reader in his or her quest for well-being. In the event that you use any of the information in this book for yourself, which is your constitutional right, the author and the publisher offer no guarantees for your subsequent experience, and assume no responsibility for your actions.

Published by
Phaelos Books
860 N McQueen Rd #1171
Chandler, AZ 85225-8104

480.275.4925
509.479.8415 fax

info@phaelos.com

www.phaelos.com

Cover and Interior Design by Adam Abraham

Front cover photo by Della Hale
Rear cover inset photo by Adam Abraham

ISBN 0-9787991-1-9

TRANSDERMAL MAGNESIUM THERAPY

Mark Sircus, Ac., OMD

with a Foreword by Daniel Reid

PHAEL**O**S
B O O K S

It is highly regrettable that the deficiency

of such an inexpensive, low-toxicity nutrient

results in diseases that cause incalculable

suffering and expense throughout the world.
                                    Dr. Steven Johnson

Magnesium is nothing short of a miracle mineral in its healing effect on a wide range of diseases as well as in its ability to rejuvenate the ageing body. We know that it is essential for many enzyme reactions, especially in regard to cellular energy production, for the health of the brain and nervous system and also for healthy teeth and bones. However, it may come as a surprise that in the form of magnesium chloride it is also an impressive infection fighter.

Walter Last

Dedication

This book is devoted first to my wife Luciana, whose love sustains and motivates me. Second to Daniel Reid, author of the *Tao of Detox* who was the first health professional who pointed me in the direction of transdermally applied magnesium chloride. Then to David Dartez who worked closely with me in opening up peoples' awareness to a totally natural form of magnesium chloride called Magnesium Oil. Thanks and appreciation to David also for the large donation of his magnesium for my research and studies which made the writing of this book possible. Then my appreciation to a host of medical scientists and doctors, especially to Dr. Carolyn Dean, author of *The Magnesium Miracle*, whose feedback and guidance you will find imbedded in the following pages. And finally, but not least, to Claudia French R.N., my research assistant and assistant director of the IMVA, who daily acts as my right and left hand in everything I write.

Table of Contents

Preface.. xiii

Foreword by Daniel Reid.. xxi

1 Magnesium..1

2 Magnesium in Modern Medicine...............................9

3 Dietary Magnesium Deficiencies...............................19

4 Magnesium and Calcium 49

5 Disease, and Preventive Care....................................63

6 Magnesium, Selenium and Zinc in Cancer Prevention...81

7 Magnesium – Antioxidant Status – Glutathione 95

8 Detoxification and Chelation 103

9 Magnesium and ATP...127

10 Magnesium and Diabetes.......................................131

11 Magnesium and Strokes..163

12 Magnesium, Violence, and Depression......................171

13 Magnesium Chloride...187

14 Magnesium Chloride vs Magnesium Sulfate..............209

15 Memory, Cognitive Function, & Periodontal Disease 213

16 Natural Influenza Protocol 217

17 Inflammation and Systemic Stress...........................229

18 Sexuality, Life and Aging 233

19 Transdermal Mineral Therapy in Sports Medicine 249

20 Menopause and Premenstrual Syndrome....................259

21 Natural Relief for Chronic Pain Sufferers...................277

22 Testing and Estimating Magnesium Levels 289

23 Warnings and Contra-indications..............................297

24 Testimonials 305

25 Magnesium — Conception to Death..........................319

Appendix..343

Index 355

Preface

The book that you hold in your hands contains information that could quite literally save your life. It certainly will extend your existence and save you and your loved ones from a considerable amount of pain. It will help you get to sleep if you are an insomniac, increase your energy levels and performance in sports if you are an athlete and help you avoid the major plagues of our time, i.e., diabetes, cancer, heart disease, neurological disorders and strokes. If you do not fall to one of these diseases your life will be extended.

For this book to achieve these goals you will have to consider an explanation for many diseases that is very different from the one promulgated by governmental and private health officials. You have to be able to entertain the idea that health authorities have gotten it seriously wrong. In reality you must be prepared to know more about health and disease than your doctor, for in their myopic obsessions with allopathic protocols and principles they have gone blind, deaf and dumb to certain medical basics.

What you are about to learn represents a great medical discovery and it all started for me in July of 2005 when I made a momentous phone call to Daniel Reid in Australia. I called to ask him, the author of *The Tao of Detox*, to tell me about a magnesium chloride product I had heard of that was applied directly onto the skin or put in one's bath. His reply was simple, "It's the best detoxification agent I know of." Researching the problems of children with autism and other neurological disorders, who are suffering from mercury poisoning and other chemical toxicities, I got excited. Very excited! My research assistant, Claudia French, RN started looking around the web and found David Dartez who was selling a natural form of magnesium chloride harvested from sea water. It is referred to as "magnesium oil" and he sent me a liter to try out. It had not yet dawned on me that my life was about to change dramatically.

At the same time I stumbled on a few other professionals in this area, especially Dr. Norman Shealy who had already been using another less natural form of magnesium chloride. I found out that Dr. Shealy had already applied for a patent to use transdermally applied magnesium chloride to raise DHEA levels and thus extend people's life spans. His research was clear: only when magnesium is applied topically onto the skin, where it passes through the fatty tissue, are DHEA levels raised. Neither oral nor intravenous magnesium administration would do this.

As director of the International Medical Veritas Association I had direct access to many doctors and scientists around the globe and I began questioning and researching. Several doctors including Dr. Garry Gordon state that it is very difficult to bring up the body's levels of magnesium through oral supplementation, and research confirms this. I began to realize that I had found something potentially very important and useful since magnesium has been reported by many reputable organizations to be deficient in over 68% of the population.

The year past I had been writing a book called *The Rising Tide of Mercury* with Dr. Rashid Buttar. Just recently he was quoted as saying, "I can now very comfortably and definitively state to you that, in my opinion, based on the evidence, every single chronic insidious disease process is related to one word: toxicity. You cannot address the issues of aging unless you address detoxification." Dr. Buttar, board certified and a diplomate in preventive medicine and clinical metal toxicology, and Vice-Chairman of the American Board of Clinical Metal Toxicology, contends that he only recently arrived at this conclusion. "Five years ago I wouldn't have said this, even a year ago I wouldn't have said it. But the more success we've had, the clearer it has become: All chronic disease is toxicity. You get rid of the toxicity and you put out the fire. You may need to rebuild afterward, but you must put the fire out. Conventional medicine is just covering your eyes so you don't see the fire."

Dr. Buttar, in my eyes, is a great clinician and I learned a lot about his approach to treating autistic children with a chelation drug called TD-DMPS. "TD" stands for transdermal so I was already familiar with this growing approach to applying medicines through the skin. It was most exciting to hear about the success he was having, bringing a good percentage of his young patients back from the shadows of autism, which most of them disappeared into after suffering damages from receiving their mercury containing vaccines. The only problem was that all of my professional life I had been a naturopath trained in Chinese medicine and acupuncture. It just went against my instincts to administer more toxic drugs to already toxic patients.

I am presently writing *Survival Medicine for the 21ˢᵗ Century* and in reality, *Transdermal Magnesium Therapy* is its main or most important chapter. If one does a search in Google for either "medical insanity" or "pharmaceutical terrorism," one will find my name. Though I am a director of an organization dedicated to medical truth, and though I have associated myself with many fine and unusually brave doctors, in general I spend most of my time writing about issues that cast a long shadow of shame on the entire medical industrial complex.

Personally I have never trusted doctors and only recently have I lost all medical respect for dentists for their blind use and support of fluoride (which can cause cancer or force you to fly to the hospital if your young child eats his or her toothpaste) and mercury bearing dental amalgam, which leaks mercury vapors in the mouth 24 hours a day. While most of today's doctors may genuinely wish to help their patients they have basically been educated and courted by pharmaceutical companies whose only interest is in lining their pockets with your money. Corporate interests are the rule of the day in the world of medicine and almost everyday more news and research comes to the forefront sustaining the fact that most pharmaceuticals do us more harm than good.

So bad is the story that we cannot even trust doctors to get their diagnoses right. Most often they are not even close. For example, some years ago much was made of two deaths from measles in Dublin. I tracked down the cases and found both children were already seriously ill before they contracted measles. When a doctor and health official says this person died of hepatitis B, measles or AIDS for that matter, the last thing we should do is believe them. Doctors easily say things that are not proven. The above is a good example. When someone dies what are they dying of? A virus, bacteria or from a severe nutritional deficiency that destroys the immune system's ability to resist? You will never hear a doctor diagnose a nutritional deficiency and yet the majority of the world is officially suffering from such deficiencies.

What you are about to read represents a monumental medical and dental discovery. We find out after all that the most basic nutritional substance from the sea is the most powerful and safe medical healing substance. Magnesium chloride as you will see, the same product when injected during a heart attack or stroke to save your life, can be used in a wide variety of disorders. I even use it as the best mouth wash you can imagine and it has saved my gums and thus my teeth from deterioration.

I am not going to tell you that it is the only nutritional/medicinal agent that you need to take but I will tell you that it is the first one you need to take and that it will enhance any other medical and dental treatment you will undergo. Personally I will not treat a patient who is not replenishing their magnesium levels for I do not enjoy failing my patients.

In this book you will see me quoting quite frequently Dr. Carolyn Dean who wrote *The Magnesium Miracle*. Her title was well chosen. Magnesium, especially in the chloride form when applied transdermally, is the *medical miracle* we have been waiting for. It is the first medicine that you should stock in your medical cabinet.

Magnesium is useful for so many things that you will need to read both this book and Dr. Dean's book to explore all its possibilities. With it you can alleviate a score of common problems like muscle pain, insomnia, migraines, menstrual pain, and depression. You can activate your vital enzyme processes and ATP production to increase your energy levels for magnesium is as much food to the body as wheat, rice, or any meat.

For all the billions and billions of dollars the pharmaceutical companies will spend on research they will never ever come up with a healing agent like magnesium chloride that comes naturally from the sea. They can huff and puff, and try to poison humanity with their toxic drugs, but they just cannot compete with the Master of Nature, with God who has created us and maintains us with the natural substances.

Just don't expect most of your doctors or dentists to either know or admit that a naturally occurring nutritional mineral can save your life in an emergency situation: at least, not yet. Magnesium chloride is one of the most potent minerals that exists and is incredibly fast acting and safe. This book calls on all doctors, dentists, naturopaths, chiropractors, nurses, acupuncturists and other health care practitioners to open their eyes to a great discovery, one that will greatly help their patients. For many though, and perhaps for the vast majority, to open to the message in this book requires humility before God and Nature that they simply do not have. Certainly the pharmaceutical companies and all their drug salesmen and women will not embrace an inexpensive safe medicine that they have to do nothing to create.

It is very sad to have to report that there is a form of terrorism afoot that is worse than anything you will read about in the daily papers. There are people around the world acting through CODEX, an organization working under the auspices of The United Nations and the World Health Organization, that want to deny you the use of vitamins and minerals.

More accurately said they want to severely limit your access to vitamin supplements, limit the quantities you can take, force you to get a medical prescription to take them, and thus drive the price for them through the roof. In the age of toxicity, where we are all being poisoned by the chemicals in the air we breathe, the water we drink, the foods we eat, and the medicines we take, we need more, not fewer, critical nutritional agents to help us detoxify our bodies and maintain our health. Magnesium chloride is one such essential nutrient, not only as a medicine but also as necessary as the air we breathe in an easily assimilated natural form.

I am an American living down in Brazil and every time a gallon of this magnesium oil arrives I am very happy. It is not widely available around the world and the obstacles for its importation are increasing because of the medical establishment's jealous husbandry over everything we take. I pay through the nose to have it shipped here and wish I had many gallons on hand to ensure my supply during the years that are coming when we all have to face hostile governments dominated by super-rich and powerful pharmaceutical companies that treat us and our children with much less respect than a farmer would treat a pig. We are all caught between an anvil and a hammer, between increasing toxicities and worsening nutritional deficiencies due to the soil depletion caused by modern farming practices and the heavy processing that our daily foods are put through.

Americans especially, and the rest of the world, take notice. If you think the $3 trillion American medical system is going to survive what is coming in economic terms you should have your head examined.

Modern medicine is going to take a great fall like Humpty Dumpty on the great wall. Not only does the system not work, killing and poisoning more people than it helps (iatrogenic death and disease), we are just not going to be able to afford medical treatments in the near future. Medical economics is another serious subject that many are already facing. More and more people just cannot afford medical treatments. Millions upon millions of citizens are losing their medical plans and or being driven into bankruptcy because of un-payable medical bills. What I am going to introduce you to in this book is the most affordable and effective medical/nutritional treatment there is and the first item on my list as a "Survival Medicine for the 21st Century."

Let us draw the line in the sand and take a position the world's medical authorities will not enjoy. In the nightmarish age of toxicity we have entered, it is the Waters of Life and magnesium chloride found in them that will help us avoid the great plagues afflicting humanity: heart disease, cancer, strokes and diabetes. It will even help children survive their vaccines as will vitamin C. In the 21st century the medical industrial complex will meet its match in a simple abundant magnesium salt from the sea.

Foreword

In this age of high-tech mechanical medicine and modern chemical pharmaceutics, the simple basic "facts of life" regarding human health and healing are all too often overlooked and forgotten. For those who are dedicated to the pursuit of real health and true healing it is therefore always cause for celebration when one of these simple basic facts is rediscovered and brought back into the light. One of the most important of these revelations in recent years is the essential role played by magnesium in almost all of the fundamental equations of human health.

The Chinese ideogram for "magnesium" consists of the symbols for "mineral" and "beautiful," hence it was known to traditional healers in China as *mei*, the "beautiful mineral," and it's importance in both preventive health care and curative therapeutics was clearly recognized. Following the guideline that food is always the best medicine, particularly in the prevention of disease and degeneration, the traditional Chinese diet contained abundant supplies of this vital mineral.

In the Western world today, particularly in America, heart disease has become one of the primary causes of premature death, and magnesium deficiency has been conclusively proven to be a major factor in all cases of heart failure. With approximately 80% of the population critically deficient in magnesium, it's small wonder that heart disease has become one of the biggest killers.

But it's not only the human heart that depends on adequate supplies of magnesium. Immune response, nerve and brain functions, blood pressure, and over 300 essential enzymatic reactions in the cells of the human body all rely on magnesium. Without adequate magnesium, most of the body's vital functions grind to a halt.

Unfortunately, most magnesium supplements on the market today are useless, for two reasons: first, they're made from the wrong form of magnesium; second, oral supplementation of magnesium is not very effective.

The form of magnesium which the human metobolic system recognizes and assimilates most readily is magnesium chloride, the same form contained in sea water, but very few nutritional supplements on the market today include this type of magnesium. And the simple secret to the proper administration and optimum assimilation of magnesium is to apply it transdermally, i.e. via the skin, not as an oral supplement.

Transdermal administration of magnesium is a quick and easy way correct chronic degeneration conditions caused by magnesium deficiency, and the simplest way to do this is to spray the surface of the skin with a fluid solution of magnesium chloride, or to soak the feet for 20 minutes in a bucket of hot water with a few ounces of magnesium chloride fluid added to it. In the integrated detoxification and regeneration healing program which my wife Snow and I offer each year at health resorts in Asia, transdermal magnesium therapy plays a key role in the form of a soothing hot bath which we refer to as a "Magnum Bath." We call our program "Renew Your Lease on Life," and the efficacy of transdermal magnesium chloride therapy for tissue detoxification and cellular regeneration have proven themselves time and again in this program.

In this book, Dr. Mark Sircus has collected together the full spectrum of essential information regarding the benefits of magnesium for human health and its practical therapeutic applications in healing. This is a book which should become required reading for all aspiring naturopathic health professionals, as well as for doctors of conventional modern medicine who are beginning to wonder why the pharmaceutical drugs they've been taught to prescribe for virtually every acute and chronic condition today not only fail to cure their patients, but often cause disastrous side-effects that lead to even worse conditions. It's also a book that everyone who wants to practice self-health care, and protect the health of one's family, should have on the shelf at home.

Daniel Reid
Byron Bay, Australia
May 2006

1

||||||||||||||||||||||||||||||||||||||||||||||||||||||||||||||||||||||||||||

Magnesium

Magnesium, atomic number 12, is an element essential for normal function of the nervous and cardiovascular systems. Pure magnesium is a silvery-white metal, which burns with a dazzling brilliance. After calcium, phosphorus, and potassium, it is the fourth most abundant mineral in cells. The two ounces or so found in the typical human body is present not as metal but as magnesium ions (positively-charged magnesium atoms found either in solution or complexed with other tissues, such as bone).

Magnesium is the second most abundant intracellular and the fourth most abundant *cation* (i.e., positively charged ion) in the body. It is an essential transmembrane and intracellular modulator of cellular electrical activity. As such, its deficiency in the body is nothing short of disastrous for cell life. Yet, this fact is not widely known.

Roughly one quarter of this magnesium is found in muscle tissue and three-fifths in bone; but less than 1% of it is found in blood serum, although that is used as the commonest indicator of magnesium status.

This blood serum magnesium can be further subdivided into free ionic, complex-bound and protein-bound portions, but the ionic portion is considered most important in measuring magnesium status, because it is physiologically active. The body works very hard to keep magnesium levels in the blood constant.

Magnesium is the single most important mineral for maintaining proper electrical balance and facilitating smooth metabolism in the cells. One of the major properties of magnesium is that of stabilizing membranes. It has a stabilizing effect not only for the cell membrane, but also for various sub-cellular organelles.

Magnesium deficiency is one of the most common nutritional problems in the industrialized world today. This deficiency is the result of agricultural practices, food preparation techniques, and dietary trends. The health implications are nothing short of catastrophic.

Magnesium is necessary for the metabolism of carbohydrates, fats and amino acids. It is essential for the functions of muscles and nerves and for the formation of bones and teeth. Generally it counteracts and regulates the influence of calcium.

There are basically two classes of minerals: micro-nutrients, which are only needed in trace amounts and macro-nutrients, of which we need fairly significant amounts. Magnesium is a mineral we need fairly large quantities of, but sadly this medical and nutritional fact has yet to be embraced in practice by general medicine.

Magnesium supplementation is grossly under utilized and under prescribed by conventional physicians. Though magnesium deficiency is common, it is usually not looked for, and therefore, not found or corrected, or factored into treatment strategies. In most industrialized countries, magnesium intake has decreased over time and is now marginal in the entire population.[1]

Magnesium deficiency can affect virtually every system of the body. Its absorption and elimination depend on a very large number of variables, at least one of which often goes awry, leading to a deficiency that can present itself with many signs and symptoms.

Dr. Carolyn Dean, author of *The Miracle of Magnesium*, has this to say. "Magnesium is very important in health and medicine. It is extremely important for the metabolism of calcium, potassium, phosphorus, zinc, copper, iron, sodium, lead, cadmium, hydrochloric acid (HCl), acetylcholine, and nitric oxide (NO), for many enzymes, for the intracellular homeostasis and for activation of thiamine and therefore, for a very wide gamut of critical body functions. Magnesium is a particularly crucial element for mediating the vital functions of the nervous and endocrine systems; it helps maintain normal muscle and nerve functions, keeps heart rhythm steady, supports a healthy immune system, and keeps bones strong. Magnesium also helps regulate blood sugar levels, promotes normal blood pressure, and is known to be involved in energy metabolism and protein synthesis. In the nucleus, more than half the magnesium is closely associated with nucleic acids and mononucleotides. Magnesium is necessary for the physical integrity of the double helix of DNA, which carries genetic information and the code for specific proteins. Enzymes are protein molecules that stimulate every chemical reaction in the body. Magnesium is required to make hundreds of these enzymes work."

Dr. Dean continues, "Of the 325 magnesium-dependent enzymes[2], the most important enzyme reaction involves the creation of energy by activating adenosine triphosphate (ATP), the fundamental energy storage molecule of the body."

ATP may be what the Chinese refer to as *qi* (pronounced, "chee"), or life force. Magnesium is required for the body to produce and store energy. Without magnesium there is no energy, no movement, no life." *Magnesium is necessary for the synthesis of various compounds that have energy-rich bonds of any type.*[3] The formation of energy-rich bonds that require Mg_2+ constitutes the necessary basis for all cellular activities. This alone establishes the critical biologic importance of magnesium. Thus fatigue is often reduced with magnesium supplementation for the many enzyme systems that require magnesium help restore normal energy levels.

The toxic effect of fluoride ions plays a key role in acute magnesium deficiency. Fluoride ion clearly interferes with the biological activity of magnesium ions. In general, fluoride-magnesium interactions decrease enzymatic activity.[4]

Dr. Dean and many other doctors and researchers are clear that "magnesium deficiency is a significant factor — often the major factor — in many severe illnesses including heart attacks and other forms of heart disease, asthma, anxiety and panic attacks, depression, fatigue, diabetes, migraines and other headaches, osteoporosis, insomnia, and most cases of muscular problems."

Dr. Steven Johnson agrees adding, "The range of pathologies associated with magnesium deficiency is staggering: hypertension (cardiovascular disease, kidney and liver damage, etc.), peroxynitrite damage (migraine, multiple sclerosis, glaucoma, Alzheimer's disease, etc.), recurrent bacterial infection due to low levels of nitric oxide in the cavities (sinuses, vagina, middle ear, lungs, throat, etc.), fungal infections due to a depressed immune system, thiamine deactivation (low gastric acid, behavioral disorders, etc.), premenstrual syndrome, calcium deficiency (osteoporosis, mood swings, etc.), tooth cavities, hearing loss, Type II diabetes, cramps, muscle weakness, impotence, aggression, fibromas, potassium deficiency (arrhythmia, hypertension, some forms of cancer), iron accumulation, etc."

Magnesium is essential in regulating central nervous system excitability thus magnesium deficiency may cause aggressive behavior,[5] depression, or suicide.[6] Magnesium calms the brain and people do not need to become severely deficient in magnesium for the brain to become hyperactive. One study[7] confirmed earlier reports that a marginal magnesium intake overexcites the brain's neurons and results in less coherence—creating cacophony rather than symphony—according to electroencephalogram (EEG) measurements.[8]

During half of the six-month study, thirteen women consumed 115 milligrams of magnesium daily—or about 40% of the Recommended Dietary Allowance (RDA). During the other half, they got 315 mg daily — a little more than the 280 mg recommended for women. After only six weeks on the marginal intake, EEG readings showed significant differences in brain function.

Magnesium exists in the body either as active ions or as inactive complexes bound to proteins or other substances.

Minerals in general rule over other nutrients because vitamins, enzymes and amino acids, as well as fats and carbohydrates, require them for activity. There are seventeen minerals that are considered essential in human nutrition and if there is a shortage of just one the balance of the entire system can be upset. A deficiency of a single mineral can negatively impact the entire chain of life, rendering other nutrients ineffective and useless.

References

1   Galan, P., Preziosi, P., Durlach, V., Valeix, P., Ribas, L., Bouzid, D., Favier, A. & Hercberg, S. (1997) Dietary magnesium intake in a French adult population. *Magnes. Res.* 10:321-328.[Medline]

2   **Enzymes of carbohydrate metabolism** glucokinase, hexokinase, galactokinase, phosphorylase phosphatase, phosphorylase kinase, phosphoglucomutase, 6-phosphofructokinase aldolase, triokinase, fructose-1,6-bisphosphatase, glucose-6-phosphatase, glucose-6-phosphate dehydrogenase, transketolase, phosphoglycerate kinase, phosphoryl glycerylmutase, enolase, pyruvate kinase, thiamine-pyrophosphate kinase, pyruvate decarboxylase, glycerokinase, glycerophosphatase, various pentoside kinases that activate B vitamins. **Enzymes of nucleic acid and protein metabolism**: RNA polymerase which allows the synthesis of RNA and especially that of messenger RNA which, associated with post-ribosomal factors of initiation and elongation and with polyamines, codes for amino acids to produce specific proteins; DNA polymerase which allows the reconstitution and recombination of DNA, ornithine carbamyl transferase, glutamine synthetase, carbamate kinase, argininosuccinate synthetase, creatine kinase, insulinase, leucine aminopeptidase which appears to be similar to hypertensinase. **Enzymes of lipid metabolism** acetylcoenzyme A synthetase, acylco A synthetase, beta-ketothiolase, diglyceride kinase, phosphatidate phosphatase, mevalonate kinase, phosphomevalonate kinase, lecithin-cholesterol-acyl transferase (LCAT).

3   The phosphoric anhydride bond that is found mainly in ATP or adenosine triphosphate, "the main fuel of life" (13), but also in GTP (guanosine triphosphate) as well as in other nucleoside triphosphates such as UTP (uridine triphosphate), CTP (cytosine triphosphate) and ITP (inosine triphosphate). It is also found in the phosphoamide bond of phosphocreatine, the phosphoenol bond of phosphoenolpyruvic acid, the mixed anhydride bond of 1,3-diphosphoglyceric acid and in the bond between an acid and a thiol group as in acyl coenzyme A or succinyl coenzyme A.

4   A Machoy-Mokrzynska. Fluoride_Magnesium Interaction. Fluoride (*J. of the International Society for Fluoride Research*), Vol. 28 No. 4; November, 1995, pp 175-177 see www.magwater.com/fl2.shtml Institute of Pharmacology and Toxicology, Pomeranian Medical Academy, Szczecin, Poland.

5   Bernard Rimland. While no patient has been cured with the vitamin B_6 and magnesium treatment, there have been many instances where remarkable improvement has been achieved. In one such case an 18-year-old autistic patient was about to be evicted from the third mental hospital in his city. Even massive amounts of drugs had no effect on him, and he was considered too violent and assaultative to be kept in the hospital. The psychiatrist tried the B_6/magnesium approach as a last resort. The young man calmed down very quickly. The psychiatrist reported at a meeting that she had recently visited the family and had found the young man to now be a pleasant and easy-going young autistic person who sang and played his guitar for her. www.autism.org/vitb6.html

6   C. M. Banki, M. Arato and C. D. Kilts. Aminergic studies and cerebrospinal fluid cations in suicide. *Annals of the New York Academy of Sciences*, Vol 487, Issue 1 221-230, Copyright © 1986 by New York Academy of Sciences

7   This is the first experimental study in which magnesium intakes were tightly controlled and EEG measurements were analyzed by computer so they could be statistically compared.

8   For more information, see: www.ars.usda.gov/is/np/fnrb/fnrb1095.htm#calm.

2

||||||||||||||||||||||||||||||||||||||||||||||||||||||||||||||||||||||||||||

Magnesium in Modern Medicine

Magnesium is nearly miraculous for the depth and scope of its application. It is not an exaggeration to say that miracles in medicine would be achieved if the overwhelming preponderance of magnesium deficiency — in adults, in adolescents, and the very young — were addressed instead of ignored. Many more lives would indubitably be saved if non-toxic medicines were favored over toxic ones. Why they are not, is another matter altogether.

This is not idle medical banter and the entire medical community will eventually have to reorient itself by putting magnesium, specifically magnesium chloride, at the top of the chart of usable medicines.

*When 1,033 hospitalized patients were studied, over 54%
were low in magnesium. What was worse is that 90% of the
doctors never even thought of ordering a magnesium test.*[1]

Journal of the AMA

Despite the fact that magnesium is almost as important for life as the air we breathe, it seems like the medical industrial complex is not too keen on the public getting enough of this precious mineral. For instance, for the past 15 years evidence has stacked up showing patients with acute coronary thrombosis improve their survival chances by 50%-82.5% when given intravenous magnesium of 32-66 mmol (1200 milligrams of magnesium equals 50 mmol) in the first 24 hours,[2] and still magnesium chloride or magnesium sulfate are not universally used in hospitals around the world. Rapid intravenous bolus doses of magnesium have been shown to instantaneously and effectively dilate the coronary collateral circulation proving to be a dramatically effective treatment of acute myocardial infarction, angina and congestive heart failure.[3]

*Magnesium is the most important mineral
to man and all living organisms.*[4]

Dr. Jerry Aikawa

The medical authorities, and certainly the pharmaceutical companies, are in a pickle with magnesium chloride. Here is a powerful medicine that is natural, non-toxic, inexpensive, and effective in a wide variety of medical situations. So what do they do? They commission a study that is essentially designed to show the opposite, in effect sabotaging medical clarity on the use of a vital, irreplaceable, safe medicine.

Specifically, one study which showed that magnesium had a worsening effect on survival, employed a far higher dose of magnesium (80 mmol) than the studies mentioned above.[5] Another study, which showed no benefit with magnesium, employed an inordinately low dose of 10 mmol in the first 24 hours.

Dr. Stephen Davies and Dr. Damien Downing, editors of *The Journal of Nutritional and Environmental Medicine*, criticized the designers of the study for clearly selecting too large a dose of intravenous magnesium, and also for giving magnesium too late and then too quickly.

"Although it would appear clear to any first year medical student that magnesium worked well for coronary thrombosis within the optimal dosage level of 30 - 70 mmol; that 10 mmol was shown to be too little, and 80 mmol had been shown to be too much."

Over 100 patients suffering from coronary heart disease were treated with intramuscular [injected] magnesium sulphate with only one death, compared to their findings in the previous year when, of 196 cases admitted and treated with routine anticoagulants, 60 died.[6]

The British Medical Journal
January 23, 1960

Because of these studies many hospitals ceased using magnesium in their treatment of acute coronary thrombosis. The scandalous decision to use an overdose of magnesium in this study is what we would expect of the profit driven pharmaceutical business and medical industrial complex that hurts more people than it helps. Iatrogenic death and disease (that is, induced by a physician or surgeon as a function of standard medical treatment or diagnostic procedures), is rampant and some of that could be avoided if magnesium were more widely used in modern medicine.

Researchers from Northwestern University School of Medicine in Chicago have determined that not having enough magnesium in our diet increases our chances of developing coronary artery disease. In a study of 2,977 men and women, researchers used ultrafast computed tomography (CT scans) of the chest to assess the participants' coronary artery calcium levels.

Measurements were taken at the start of the study — when the participants were 18 to 30 years old — and again 15 years later. The study concluded that dietary magnesium intake was inversely related to coronary artery calcium levels. In other words, the lower the magnesium intake through diet, the higher the calcium levels in the coronary system. Coronary artery calcium is considered an indicator of the blocked-artery disease known as atherosclerosis.

Science is starting to realize magnesium's role as the number one preventative agent for the major plagues of modern man. In two huge long-term studies it was concluded that people who consumed the most magnesium in their diet were least likely to develop Type II diabetes. This is according to a report in the January 2006 issue of the journal *Diabetes Care*. Until this groundbreaking study, very few large-scale research efforts have directly examined the long-term effects of dietary magnesium on diabetes. Dr. Simin Liu of the Harvard Medical School and School of Public Health in Boston says, "Our studies provided some direct evidence that greater intake of dietary magnesium may have a long-term protective effect on lowering risk."

Considering some of the basic research already published it is sometimes frustrating to witness the reticence within the medical community shown toward using magnesium as a primary medicine.

Dr. Russell Blaylock describes his own experience with this, telling how his own brother fell victim to cancer, and how the lack of proper treatment led to a death that could have been prevented.

"I asked the doctor in charge of his respiratory care to add vitamins and magnesium to his IV. While he promised he would, he didn't. When I asked his doctor why the magnesium had not been added to his IV, word was sent back to me through the nurse that she had never heard of using magnesium. I sent copies of selected articles showing the immense value of magnesium on pulmonary and cardiovascular function. Still there was no response from the doctor."[7]

Imagine this scenario happening hundreds, thousands, and even million of times again, over the past fifty years. While we can't do anything for those who have needlessly departed, we can certainly do something who are presently, and will be needlessly put at risk, save for the simple application of this essential natural nutrient.

Magnesium deficiency commonly occurs in critical illness and correlates with a higher mortality and worse clinical outcome in intensive care units. Studies are now underway that have emergency crew personnel authorized to administer IV magnesium immediately in the ambulance. Preliminary trials found "promising" effects of $MgSO_4$ (magnesium sulfate) on stroke victims, if given early enough, before getting to emergency rooms.[8] Magnesium infusion in patients with acute myocardial infarction (four grams of $MgSO_4$ during the first three days) reduced the incidences of arrhythmias, death and the size of infarction. Another study showed reduction of mortality with infusion of 10 grams of $MgSO_4$ in 24 hours.[9]

Dr. Sarah Mayhill, a British doctor working for the National Health Service says, "In fact it is partly this effect which is taken advantage of in the treatment of acute myocardial infarction or acute stroke. In both these conditions there is a local obstruction of blood supply. I use IV magnesium (2-5mls of 50%) as a bolus (i.e., dose) to treat both these conditions, often with dramatic effects.

With acute myocardial infarctions there is often immediate pain relief, as either the obstruction is relieved or good collateral circulation restored. Furthermore, magnesium is antiarrhythmic. Trials with magnesium have clearly demonstrated benefit and magnesium is used as a front line drug in many hospitals. In acute stroke, function can be restored within a few minutes: most satisfying. However, if there is a possibility that the stroke is hemorrhagic (about 15% of cases) then magnesium should not be used."

Intravenous magnesium is safe and effective in acute severe asthma and is commonly used by emergency medical personnel.

Magnesium has many known indications in anesthesiology and intensive care, and new studies are beginning to suggest its use in many other areas of medicine as well. For instance two studies have suggested magnesium's role in the treatment of acute migraine. Mauskop et al[10] demonstrated relief of headache within 15 minutes of intravenous magnesium in 32 of 40 patients with migraine, cluster headache, or tension headache.

"Not all headaches are produced by mineral imbalances, but we now know that fifty to 60% of migraines are magnesium-linked. And that's probably why no prescription therapy on the market successfully treats headaches across the board. They're simply not treating the cause," says Dr. Burton M. Altura, professor of physiology and medicine at the State University of New York Health Science Center at Brooklyn.

"Of the seventeen people we've treated with magnesium, thirteen have had complete improvement," says Dr. Herbert C. Mansmann, Jr., professor of pediatrics and associate professor of medicine at Jefferson Medical College in Philadelphia.[11]

When used correctly, magnesium chloride is a weapon against infectious diseases. Between its power to stimulate white blood cells and glutathione production, and its basic role in producing energy, we have a heavyweight non-toxic medicine that can be used without a prescription. This is going to be very important as antibiotics, which do nothing to enhance the body's ability to protect and heal itself, become increasingly ineffective.

An example of this phenomenon is where an increasing number of young, otherwise healthy Americans who are being stricken by the bacterial infection known as *Clostridium difficile* — or *C. diff* — which appears to be spreading rapidly around the country and causing unusually severe, sometimes fatal illness.

> *It's a new phenomenon. It's just emerging. We're very concerned. We know it's happening, but we're really not sure why it's happening or where this is going.*
> Centers for Disease Control

The infection has long been common in hospital patients taking antibiotics for other reasons. As the drugs kill off other bacteria in the digestive system, the *C. diff* microbe can proliferate.

Hospitals might be forced to use magnesium chloride or just watch as more and more die from their refusal to step outside their medical boxes and use something that can safely help deal with this and other medical situations.

Magnesium chloride, when concentrated, is a powerful universal medicine that we can turn to in many clinical situations, including common influenza and the "dreaded" bird flu, especially when used in conjunction with vitamin C. This is an exciting medical discovery. The same pure natural substance used in emergency rooms to save people's lives has a dramatic effect on cell life and is safer to use than aspirin. Effective in a much broader sense than vitamin C, magnesium chloride is a medicine that helps doctors to fulfill their primary mission and purpose.

"Magnesium is necessary for the normal function of over 300 enzyme systems, for muscle relaxation, immune function, cardiac function, clotting, nerve conduction etc. Indeed I cannot think of a bodily department in which magnesium is not essential. It prevents heart disease, cancer, high blood pressure, kidney stones and improves energy, sleep etc." reports Dr. Sarah Mayhill.

Sarah Mayhill, M.D., works both for the National Health Service (UK) and in private practice. About 10% of her NHS patients suffer from Chronic Fatigue Syndrome (CFS) and approximately 70% in her private practice have it. Dr. Mayhill is a medical advisor to Action for ME, a national support organization in the United Kingdom for ME/CFS sufferers. She is also the Honorary Secretary of the British Society for Allergy Environmental and Nutritional Medicine.

Dr. Mayhill has written extensively about CFS over the years, covering all aspects of the disease from diagnosis to causal theories to treatments. This excerpt was adapted from her book *Diagnosing and Treating Chronic Fatigue Syndrome*, and is used with permission of the author.

"Like two diverging paths, it appears that the more we learn about the benefits of magnesium the more we uncover about the side effects of prescription drugs," says Dr. Carolyn Dean, author of *The Miracle of Magnesium*.

Magnesium chloride is a versatile and absolutely essential mineral that should be in every body. Where physiological magnesium levels are insufficient, it should be near, somewhere in each household, and in every medicine cabinet.

It boosts almost all aspects of cell physiology and is what you want around in the event of a heart attack or stroke, although if your magnesium intake is sufficient, your odds of having a heart attack or stroke are greatly diminished.

Magnesium chloride treatments address systemic nutritional deficiencies, act to improve the function of our cells and immune system, and help protect cells from oxidative damage. It is a systemic medicine as well as a local one that brings new life and energy to the cells wherever it is applied.

Hundreds of billions of dollars and millions of lives would be saved if magnesium was supplemented and used widely as a food, or as a medicine. Magnesium chloride is a basic mineral nutrient that can be used orally, intravenously, and transdermally (i.e., absorbed into the bloodstream directly through the skin). While this book will discuss all of these treatment modalities, the many advantages of the transdermal approach will become crystal clear.

References

1   June 13, 1990

2   *J Nutr Environ Med*, 1999;9:513

3   S. E. BROWNE. Review of 34 years of experience. *Journal of Nutritional Medicine* (1994) 4, 169-177. www.mgwater.com/browne02.shtml

4   Aikawa LK, *Magnesium: Its Biological Significance*, CRC Press, Boca Raton, Fl, 1981

5   *European Heart J*, 1991;12:12158

6   Rodale J.I., Taub, Harald J. *Magnesium, the nutrient that could change your life.* Pyramid Books. New York.

7   *How Modern Medicine Killed My Brother.* www.mercola.com/2004/nov/24/modern_medicine.htm

8   See: www.fastmag.info/sci_bkg.htm and www.fastmag.info/index.htm

9   Faintuch JJ, Menezes MS. Magnesium and myocardial infarction. Brazilian aspects. Clinicia Geral do Hospital das Clinicas, Faculdade da Universidade de Sao Paulo. Rev Hosp Clin Fac Med Sao Paulo. 1997 Nov-Dec;52(6):333-6. Most of the brazilian's territory is poor in magnesium (Mg) and an evaluation of urinary Mg indicated very low concentration of this cation in a normal population sample. The study of the behavior of plasmatic Mg in the acute phase of uncomplicated myocardial infarction permitted the following conclusions; a) during the first three days of the clinical course there is significant hypomagnesemia; b) magnesemia rises progressively during the three days of infarction, without however reaching normal levels. The lymphocytic magnesium also show the same behavior.

10   Mauskop A, Altura BT, Cracco RQ, et al. Intravenous magnesium sulfate rapidly alleviates headaches of various types. Headache 1996;36:154–60.[Medline]

11   See: www.mgwater.com/prev1801.shtml

3

|||||||||||||||||||||||||||||||||||||||||||||||||||||||||||||||||||||

Dietary Magnesium Deficiencies

Studies show that as many as half of all Americans do not consume enough magnesium. Magnesium deficits have been tied to allergies, asthma, attention deficit disorder, anxiety, heart disease, muscle cramps and other conditions.[1]

Massachusetts Institute of Technology

The latest government study shows a staggering 68% of Americans do not consume the recommended daily intake of magnesium. Even more frightening are data from this study showing that 19% of Americans do not consume even half of the government's recommended daily intake of magnesium.[2] The implications of this behavior stand before us as the current state of *ill-health* in America.

One of the great challenges in medicine today is to understand the complexity of causes that lead to the breakdown of health and the formation of serious disease. There are so many factors that simultaneously impinge on our physical systems that it is truly a daunting task to ascertain what is causing what. During this past century the physical environment that surrounds us has gotten incredibly toxic to the extent that *most popular foods we consume diminish rather than enhance the body's health.*

There are people and organizations that hide behind this complexity of causes, thus making it difficult to prove that any large scale, or worse, intentional harm is being done. Yet, answers to the cause of such ailments as autism, sudden infant death syndrome (SIDS), why some children mysteriously die after being vaccinated, officially remain unrequited.

Let thy food be thy medicine, and thy medicine be thy food.
Hippocrates, 400 B.C.

A few years ago I wrote a *Tale of Two Hammers* about the situation in Africa where populations were being decimated because mass vaccine programs were being administered to malnourished populations whose immune systems were already compromised. Little did I dream then of a similar situation in the west with the majority of the population being malnourished in magnesium.

Food contamination is a growing problem and now an acknowledged risk to young children and adults alike. It does not take much to see that human well being has been breached by the air we breathe, the water we drink, by medicines and vaccines administered to us (medicines and vaccines we are encouraged to believe that we *need* in order to stay healthy), by mercury fillings in our mouths, and clearly by many chemical additives, some of which are "undocumented", that are routinely placed in our food.

At least 2,800 substances have been recognized as food additives by the U.S. Food and Drug Administration. These are used to make foods more attractive, to make foods tastier, and to increase the grocery shelf life.

The Pesticide Action Network's (UK) analysis reveals a diverse cocktail of chemicals in food. "Mostly, but not always, below legal limit, 65% of them are recognized hazards to health: 35% are suspected cancer-causing chemicals, 12% are hormone-disrupting chemicals, and 41% are acutely toxic."

Because magnesium is so important for the removal of toxic substances from the body its lack makes us even more vulnerable to food contamination. According to Dr. Carolyn Dean if you have a magnesium deficiency and regularly use aspartame, the toxicity is magnified and can result in headaches and migraines.

More and more people are becoming aware of the silent chemical mayhem being waged against us and our children, but few are aware of the shrinking values of actual vitamins, minerals and proteins in the food we eat.

On one side we are literally being poisoned by added chemical ingredients, and on the other we are actually getting less of the very nutrition necessary to resist the effects of all the different toxins we are ingesting. Then, on top of everything else, our systems have to navigate through further deficiencies brought on by the cumulative effect of allopathic drugs like antibiotics that are prescribed by doctors, or otherwise habitually sought by us, way too often.

On those occasions that we take positive action and use chelators (drugs used to remove heavy metals), important minerals are reduced even further.

Drug/Substance	Nutrients Depleted
Antibiotics	vitamin A, B-12, C, E, K, Biotin, Calcium, Iron, **Magnesium**, Potassium
Chelators	Copper, Iron, **Magnesium**, Zinc
Anticonvulsants	vitamin B-2, B-12, C, F, K, Folic Acid, Calcium, **Magnesium**
Antidiabetics (Oral)	vitamin B-2, B-12, C, D, Folic Acid
Antihistamines	vitamin C
Aspirin	Calcium, Folic Acid, Iron, Potassium, C, B Complex

Dr. Matthias Rath says that, "Almost all the prescription drugs currently taken by millions of people lead to a gradual depletion of vitamins and other essential cellular nutrients in the body. Drugs are generally synthetic, non-natural substances that we absorb in our bodies. Our bodies recognize these synthetic drugs as 'toxic,' just like any other non-natural substance. Thus, all synthetic drugs have to be 'detoxified' by the liver in order to eliminate them from our bodies.

"This detoxification process requires magnesium and vitamin C and other cellular nutrients as cofactors. Many of these essential nutrients are used up in biological (enzymatic) reactions during this detoxification process. One of the most common ways for eliminating drugs from our bodies is called *hydroxylation*. The strongest 'hydroxylating agent' in our bodies is vitamin C, which is literally destroyed during this detoxification process. Thus, long-term use of many synthetic prescription drugs leads to chronic vitamin depletion in the body, a form of early scurvy and the onset of cardiovascular disease."

*Micronutrient content of the average diet
in industrialized countries is declining.*

Cheryl Long and Lynn Keiley writing for *Mother Earth News*[3] tell us that "American agribusiness is producing more food than ever before, but the evidence is building that the vitamins and minerals in that food are declining. For example, eggs from free-range hens contain up to 30% more vitamin E, 50% more folic acid and 30% more vitamin B_{12} than factory eggs.

"Most of our food now comes from large-scale producers who rely on chemical fertilizers, pesticides and animal drugs, and inhumane confinement animal production. In agribusiness, the main emphasis is on getting the highest possible yields and profits; nutrient content (and flavor) are, at best, second thoughts. This shift in production methods is clearly giving us less nutritious eggs and meat. Beef from cattle raised in feedlots on growth hormones and high-grain diets has lower levels of vitamins E, A, D and beta carotene, and twice as much fat, as grass-fed beef."

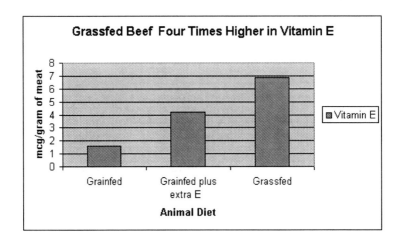

Health writer Jo Robinson has done groundbreaking work on this subject[4] making us critically aware of the importance of the conditions in which our crops, meat and dairy are raised.

Data from: Smith, G.C. "Dietary supplementation of vitamin E to cattle to improve shelf life and case life of beef for domestic and international markets." Colorado State University, Fort Collins, Colorado

We are not getting the minerals we need because modem agricultural methods, including widespread use of N P K fertilizer, over farming, loss of protective ground cover and trees, and lack of humus have made soils vulnerable to erosion. The result is a reduced nutrient content of crops.

N P K fertilizer is highly acidic. It disrupts the pH (acid/alkaline) balance of the soil, as does acid rain. Acid conditions destroy soil microorganisms whose job it is to transmute soil minerals into a form that is usable by plants. In the absence of these microbes, these minerals become locked up, unavailable to the plant.

Stimulated by the N P K fertilizer, the plant grows, but is deficient in vital trace minerals. In the absence of trace minerals, plants take up heavy metals (such as aluminum, mercury and lead) from the soil. Between 1950 and 1975, the calcium content in one cup of rice dropped 21%, and iron fell by 28.6%.

*When trace minerals are scarce in plant
bodies they become scarce in human bodies.*

Dr. Scott Whitaker, in his book *MediSin*, tells us how unfortunate it is that the modern day farmer has been persuaded to use monoculture, artificial fertilization, pesticides, and herbicides. "The end result of our domestic food production has been 'quantity' rather than 'quality'. The human body can thrive on fruits and vegetables that are grown on vital rich soil but not on soil that is artificially pumped up with chemicals." Thus today hardly anyone can eat enough fruits and vegetables to supply his or her body with the mineral salts required for good health.

*It is crucial that doctors and parents recognize that **from poor soil comes poor food,** deficient in minerals and vitamins*

Dr. Nan Kathryn Fuchs, author of *The Nutrition Detective*, says that, "Our diets today are very different from those of our ancestors though our bodies remain similar. Thousands of years ago, our ancestors ate foods high in magnesium and low in calcium. Because calcium supplies were scarce and the need for this vital mineral was great, it was effectively stored by the body. Magnesium, on the other hand, was abundant and readily available, in the form of nuts, seeds, grains, and vegetables, and did not need to be stored internally. Our bodies still retain calcium and not magnesium although we tend to eat much more calcium (in the form of dairy products) than our ancestors.

"In addition, our sugar and alcohol consumption is higher than theirs, and both sugar and alcohol increase magnesium excretion through the urine. Our grains, originally high in magnesium, have been refined, which means that magnesium is lost in the refining process. The quality of our soil has deteriorated as well, due to the use of fertilizers that contain large amounts of potassium, (which is) a magnesium antagonist. This results in foods lower in magnesium than ever before."

We need an average of 200 milligrams more magnesium than we get from the average diet.

Dr. Mildred Seelig
President of the American College of Nutrition

Imbedded into allopathic medicine are starvation nutritional protocols. The United Nation's Codex Alimentarius commission actually is staffed by people in the medical community whose main goal is to *block* our access to vitamins and minerals within the therapeutic range, as well as to all the most innovative dietary supplements, especially food based products, and even to organic food. If they have their way we will only be able to buy at highly inflated prices — with a doctor's prescription — low levels of vitamins and minerals.

Allopathic philosophy ignores the idea or concept of nutritional deficiency. There is no general awareness when a person's disease is caused by a deficiency in a vital mineral like magnesium. Western medical science missed the boat completely when it came to the declining values of magnesium in food. Imagine some huge eye going blind to what modern farming and food processing has been doing to the nutritional values of food. It means that no attention has been paid to the slowly increasing levels of malnutrition, and chronic illness in populations. In the first world we now find the majority of obese people are actually malnourished in essential minerals and vitamins.

The food supply has been steadily becoming magnesium-poor since 1909.[5]

1909 intake	408 mg/day
1949 intake	368 mg/day
1980 intake	349 mg/day
1985 intake	323 mg/day (men)
1985 intake	228 mg/day (women)

There has been a steep decline of dietary magnesium in the United States, from a high of almost 500 mg/day at the turn of the last century to barely 175-225 mg/day today.[6] The National Academy of Sciences also has determined that most Americans are magnesium deficient. Their calculations are that men obtain only about 80% of their daily needs with women fairing even worse obtaining about 70% of their needs.[7]

The magnesium content of refined foods is usually very low. Magnesium is a fairly soluble mineral, which is why boiling vegetables can result in significant losses; in cereals and grains, it tends to be concentrated in the germ and bran, which explains why white refined grains contain relatively little magnesium by comparison with their unrefined counterparts.

Whole-wheat bread, for example, has twice as much magnesium as white bread because the magnesium-rich germ and bran are removed when white flour is processed. Magnesium deficiency is more likely in those who eat a processed-food diet; in people who cook or boil all foods, especially vegetables, and in people who eat food grown in magnesium-deficient soil, where synthetic fertilizers containing no magnesium are often used.

Deficiency is also more common when magnesium absorption is decreased, such as after burns, serious injuries, or surgery and in patients with diabetes, liver disease, or intestinal mal-absorption problems. Also deficiencies develop when magnesium elimination is increased, which it is in people who use alcohol, caffeine, or excess sugar, or who take diuretics or birth control pills. We can add to this list vaccines because they offer a traumatic insult to the body that have to be defended against and that defense gobbles up both magnesium and vitamin C.

A major hidden precipitating factor in magnesium deficiency is what is called magnesium wasting that increases with increasing metabolic acidosis. New research suggests that the acid-base status affects renal magnesium losses, irrespective of magnesium intake. Magnesium deficiency could thus, apart from an insufficient intake, partly be caused by the acid load in the body.[7] This is a big problem considering the dramatic shift in past decades to processed acid forming foods.

Other drugs that cause loss of body magnesium:

Cocaine	Diuretics
Beta-adrenergic agonists (for asthma)	Thiazide
Corticosteroids (CS) (for asthma)	Phosphates (found in cola drinks)
Theophylline (for asthma)	Nicotine
Insulin	

The nutrient content of foods can no longer be relied upon. The effects of stress, intense physical activity, or the use of certain medications cause magnesium deficiency.

Because magnesium in certain forms is not easily absorbed and because no classical symptoms exist that point to magnesium's causal role in disease, the problem of its deficiency is readily masked. Many are the conditions that reduce total body magnesium and increase magnesium requirements. With nutritional values declining quickly and chemical toxicity in our bodies rising rapidly we and our children are caught between a rock and a hard place.

Data indicate that subsets of the population may be unusually susceptible to the toxic effects of fluoride and its compounds. These populations include the elderly, people with magnesium deficiency, and people with cardiovascular and kidney problems.[8]

Several studies have reported that increasing calcium in the diet significantly reduces the absorption of magnesium. In addition, diarrhea, extreme athletic physical training, sodas (especially cola type sodas, both diet and regular), sodium (high salt intake), stress (physical and mental—anything that activates a person's fight or flight reaction), and intense sweating all diminish magnesium levels.

Magnesium deficiency at a cellular level where it counts is not easy to diagnose, as serum magnesium levels do not correlate to muscle or cellular magnesium levels. Instead of trying difficult tissue magnesium analysis to find out if your health problems may be due to low magnesium levels, it is much easier and more effective just to take more magnesium and see what happens. Caution is necessary only in cases of renal failure.

Table 1: The magnesium content of common foods

Food	MAGNESIUM Content (milligrams per 100g)
Pumpkin seeds (roasted)	532
Almonds	300
Brazil nuts	225
Sesame seeds	200
Peanuts (roasted, salted)	183
Walnuts	158
Rice (whole grain brown)	110
Wholemeal bread	85
Spinach	80
Cooked beans	40
Broccoli	30
Banana	29
Potato (baked)	25
White bread	20
Yoghurt (plain, low fat)	17
Milk	10
Rice (white)	6
Cornflakes	6 ('Frosties' or 'Honeynut')
Apple	4
Honey	0.6

Source; USDA Nutrient Database

Green vegetables such as spinach are good sources of magnesium because the center of the chlorophyll molecule (which gives these vegetables their color) contains magnesium. Since 1981, Life Extension[9] has recommended high-potency magnesium supplements, because magnesium is the most deficient mineral in the American diet. In the early 1980s, the Life Extension Foundation was criticized by mainstream doctors for recommending high doses of magnesium relative to calcium. They even had their magnesium supplements seized by the FDA because they presented evidence that this mineral could help prevent heart attack.

An excess of a toxic metal and/or a relative deficiency of a nutritional element can be found as significant contributors to every disease.

Dr. Garry Gordon

William Faloon from Life Extension says, "With all the research linking low magnesium intake with high cardiovascular risks, this low-cost mineral would appear to be a simple way to counter today's heart attack and stroke epidemic. Unfortunately, magnesium is so cheap that virtually no one is promoting it as a life saving mineral."

There is no substitute for magnesium; it's as close as a metal comes to being as necessary as air.

Intravenous, Transdermal, and Oral Magnesium Mineral Therapy

"Magnesium is poorly absorbed orally. That is why I start off with injections. By injecting magnesium I can guarantee 100% to bring the levels up. I cannot guarantee to do this with oral magnesium," says Dr. Sarah Mayhill who continues with, "Treating magnesium deficiency is the most difficult deficiency to correct. In evolutionary terms, magnesium was abundant in the diet and therefore no good mechanisms to conserve magnesium evolved. It appears to be poorly absorbed and easily excreted even by normal people."

The problem with oral magnesium is that all magnesium compounds are potentially laxative. And there is good evidence that magnesium absorption depends upon the mineral remaining in the intestine at least 12 hours. If intestinal transit time is less than 12 hours, magnesium absorption is impaired, and this is the case when high doses of oral magnesium are administered. Thus it is very difficult to administer what would be considered medicinal doses orally.

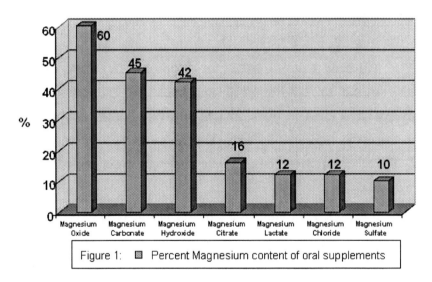

Figure 1:   ▦ Percent Magnesium content of oral supplements

Of the many forms of oral magnesium,[10] perhaps one is more easily utilized then the other. Oral magnesium chloride is well tolerated, gets absorbed very quickly, and is inexpensive. Magnesium chloride hexahydrate ($MgClH_6$) can be purchased chemically pure from most chemical supply houses without a prescription. One of the disadvantages of oral magnesium compositions that are currently available is that they do not control the rate of release. One reason for the inefficiency is that the magnesium is released into the upper gastrointestinal tract where it reacts with other substances such as calcium. These reactions reduce magnesium's rate of absorption.

Many things affect magnesium absorption from the gut. A number of factors can prevent the uptake of minerals, even when they are available in our food. For example, the glandular system that regulates the messages sent to the intestinal mucosa require plentiful fat-soluble vitamins in the diet to work properly. Likewise, the intestinal mucosa (mucous membrane) requires fat-soluble vitamins and adequate dietary cholesterol to maintain proper integrity so that it passes only those nutrients the body needs, while at the same time keeping out toxins and large, undigested proteins that can cause allergic reactions.

Minerals may "compete" for receptor sites. Excess calcium may impede the absorption of manganese, for example. Lack of hydrochloric acid in the stomach, an over-alkaline environment in the upper intestine or deficiencies in certain enzymes, vitamin C and other nutrients may prevent chelates[11] from releasing their minerals.

Finally, strong chelating substances, such as phytic acid in grains, oxalic acid in green leafy vegetables and tannins in tea, may bind with ionized minerals in the digestive tract and prevent them from being absorbed.

Most drugs will adversely affect how magnesium taken orally is absorbed or how quickly it will be excreted. Consider the antipsychotic drugs used to treat children with autism for behavior control. *Zyprexa*, *Risperdal*, and others can cause hyperglycemia, which causes increased excretion of orally administered magnesium.

While drugs bind with magnesium, diminishing its availability in the body, the threat of depletion doesn't end there. Two cans of soda per day (all of which contain phosphates) can also bind up significant amounts of magnesium. The phosphorus contained in these "harmless" beverages prevents the absorption of magnesium ions in the GI tract. In the same way, magnesium binds with aspartame, which is a major ingredient in diet sodas.

Clearly, in this day and age, magnesium supplementation is actually crucial for everyone today. However, the method of supplementation is critical in terms of effective body utilization.

Magnesium is absorbed primarily in the distal small intestines or colon. Active uptake is required involving various transport systems such as the vitamin D-sensitive transport system. Since magnesium is not passively absorbed it demonstrates saturable absorption resulting in variable bioavailability averaging 35-40% of administered dose, even under the best conditions of intestinal health.

Magnesium levels in the body, presence of calcium, phosphate, phytate, and protein can affect rate of absorption. These and other conditions make oral magnesium supplements intake chancy and inefficient compared to the new transdermal magnesium chloride mineral therapy that this book introduces.

The health status of the digestive system and the kidneys significantly influence magnesium levels.

Magnesium is absorbed in the intestines and then transported through the bloodstream to cells and tissues. Approximately one-third to one-half of dietary magnesium is absorbed into the body.[12]

Gastrointestinal disorders that impair absorption such as Crohn's disease can limit the body's ability to absorb magnesium.

"When people are ill, faced with magnesium deficiency and poor digestion, what do you think the odds are of fixing that problem with oral magnesium supplementation and digestive enzymes alone?" asks Dr. Ronald Hoffman.

In his clinic Dr. Hoffman carefully measures magnesium and found that many patients with low magnesium who take oral supplements alone do not normalize.

Dr. Mildred Seelig, renowned researcher of magnesium, predicts that through oral supplementation, it would take six months to normalize magnesium levels in a woman who is magnesium deficient. Transdermal magnesium therapy speeds up the process of nutrient replenishment in much the same way as intravenous methods.

For children with neurological disorders or asthma,
transdermal magnesium is like an oxygen mask.

Dr. Mayhill tells us, "One injection of 2mls of 50% magnesium sulphate (1gm $MgSO_4$, or 100mgs elemental Mg or 4 millimols) will usually keep levels up for two weeks. However, some people need them more often. By the third week, levels will usually have fallen again.

This is the only method that has worked for some, but it is tedious to have to keep injecting. Furthermore, because a concentrated solution is being injected the procedure, described below, is a painful one.

Intramuscular magnesium injection

Intramuscular magnesium injection is best given at room temperature or blood heat, either into the triceps or deltoid. It should be administered slowly, over a one to two minute period. I usually use an orange needle, at least one inch in length, to penetrate deep into the muscle.

Magnesium is a powerful vasodilator. Even if one takes care to check the tip of the needle is not in a vein, sometimes there is such a powerful local vasodilatation that the vessels open up and an intravenous injection is inadvertently given. This does not matter much, except that the patient develops a generalized vasodilatation, feels hot and alarmed, goes red and may faint (if upright)."

Intravenous Magnesium

According to Dr. Norman Shealy the most rapid restoration of intracellular magnesium is accomplished with intravenous replacement. For most patients ten (10) shots, given over a two-week period, are adequate. Depending upon the patient's weight and general status, Dr. Shealy gives either 1 or 2 grams of magnesium chloride intravenously over a 30 to 60 minute period:

Magnesium I	Magnesium II
• 250 cc of 0.9% Sodium Chloride	• 250 cc of 0.9% Sodium Chloride
• 1 gram Magnesium Chloride	• 2 grams Magnesium Chloride
• 500 mg Calcium Chloride	• 1 gram Calcium Chloride
• 100 mg. Pyridoxine (B_6)	• 100 mg. Pyridoxine (B_6)
• 1 gram DexPanthenol (B_5)	• 1 gram DexPanthenol (B_5)
• 1000 mcg Cyanocobalamin (B_{12})	• 1000 mcg Cyanocobalamin (B_{12})
• 6 grams vitamin C	• 6 grams vitamin C

Therapy with magnesium is rapid acting, has a safe toxic-therapeutic ratio, is easy to administer and titrate.[13] Magnesium has minimal side effects in usual therapeutic doses and has a large therapeutic index. All of this means it is so useful that it is just negligent to not use it. There is no medicine like magnesium chloride when it comes to the effect it has on the life of cells.

Though injection is the quickest way of restoring normal blood and tissue levels of magnesium, the injections, while giving benefit, are just too painful to be considered for children and for long term use in adults. They are also relatively expensive because they have to be administered by a doctor.

Introducing transdermal magnesium therapy

Transdermal magnesium, which involves absorption directly through the skin, is inexpensive, safe, a do-it-yourself at home technique that can easily replace uncomfortable injections in anything other than emergency room situations.

By using what is called "magnesium oil," either topically or in a soak, massive amounts of magnesium can be absorbed naturally, by our bodies. Body pains can be eliminated quickly in a strong soak or through direct application to the skin. Magnesium oil is made up of approximately 31 to 35% magnesium chloride, derived from natural sources and is both moisturizing and hydrating.

Transdermal application of magnesium is actually superior to oral supplements in many ways and is the best practical way magnesium can be used as a medicine besides by direct injection. Used transdermally or intravenously we have a potent natural substance that penetrates the cells with stunning result on cell biochemistry. This includes the body's ability to heal, overall energy production (ATP), skin integrity, cardiac health, diabetes prevention, pain management, a calming effect on the nervous system, sleep improvement, lowering of blood pressure, are among the general uses magnesium chloride can be put to.

The studies coming out every day provide more evidence of the need to supply adequate magnesium to people of all ages, and in a form that will be easily absorbed.

What a few can do with intravenous magnesium injections, everyone can do with transdermal magnesium.

Dr. Norman Shealy has done studies on transdermal magnesium chloride mineral therapy where individuals sprayed a solution of magnesium chloride over the entire body once daily for a month and did a 20 minute foot soak in magnesium chloride also once daily. Dr. Shealy recruited 16 individuals with low intracellular magnesium levels; subjects had a baseline Intracellular Magnesium Test documenting their deficiency and another post-Intracellular Magnesium Test after 1 month of daily soaks and spraying were analyzed. The results: 12 of 16 patients, 75%, had significant improvements in intracellular magnesium levels after only four weeks of foot soaking and skin spraying.

Typical Results: Test results before and after 4 weeks of foot soaks:

Electrolyte Name	Foot Soaking		Reference Range
	Before Soaking	After Soaking	
	(mEq/l)	(mEq/l)	(mEq/l)
Magnesium	31.4	41.2	33.9 - 41.9
Calcium	7.5	4.8	3.2 - 5.0
Potassium	132.2	124.5	80.0 - 240.0
Sodium	3.4	4.1	3.8 - 5.8

Chloride	3.2	3.4	3.4 - 6.0
Phosphorus	22.2	17.6	14.2 - 17.0
Phosphorus/Calcium	3.0	3.7	3.5 - 4.3
Magnesium/Calcium	4.2	8.6	7.8 - 10.9
Magnesium/Phosphorus	1.4	2.3	1.8 - 3.0
Potassium/Calcium	17.6	26.1	25.8 - 52.4
Potassium/Magnesium	4.2	3.0	2.4 - 4.6
Potassium/Sodium	39.1	30.5	21.5 - 44.6

Intravenous as well as transdermal administration of magnesium bypass processing by the liver. Both transdermal and intravenous therapy create "tissue saturation", the ability to get the nutrients where we want them, directly in the circulation, where they can reach body tissues at high doses, without loss.

Transdermal magnesium lotions deliver high levels of magnesium directly through the skin to the cellular level, bypassing common intestinal and kidney symptoms associated with oral use. Magnesium chloride has a major advantage over magnesium sulfate because it is hygroscopic and will attract water to it, thus keeping it wet on the skin and vastly more likely to be absorbed, while magnesium sulfate simply "dries" and becomes "powdery".

Magnesium oil feels "oily" on the skin. The biggest benefit of topical, transdermal magnesium chloride administration, is that the intestines are not adversely impacted as they might by ingesting large doses orally.

The correction of magnesium deficit is a top priority for clinicians. When magnesium chloride is understood properly, as the basic and *essential* medicine it is, it will be prescribed to all patients as a foundation and support for all other therapeutic and pharmaceutical interventions.

The same medicine that can be used as a treatment to limit myocardial damage in myocardial infarction[14] can be used safely for a broad range of problems that healthcare practitioners see everyday.

Dr. Walt Stoll says, "Magnesium deficiency inhibits the body's ability to absorb magnesium. This is an idiosyncrasy of magnesium. Once the intracellular level gets low enough to cause symptoms, in some people, the intestinal lining loses its ability to absorb magnesium efficiently. The purpose of intravenous magnesium treatment is to get the body over that hump so that it can be absorbed orally again." The same could be said about magnesium applied through transdermal/topical means.

In summary, magnesium is a safe and simple intervention and should be the first thing doctors recommend to their patients. Transdermal mineral therapy with magnesium chloride is the most powerful, relatively safe medical intervention we have to care for many of our patients needs. With the simple application of an oily solution on the skin or used in baths we can easily have our patients take up their magnesium to healthier levels. With patients who are deficient in magnesium (the great majority of patients are magnesium deficient) expect dramatic improvements in a broad range of conditions.

Magnesium Treatments

Magnesium chloride is so safe and effective we have to wonder why it has been ignored by allopathic medicine. One explanation was offered as early as 1943 by Dr. A Neveu, "Official Medicine saw in magnesium chloride Therapy a threat to its new and growing business: vaccinations." In the 1940's Dr. Neveu used the magnesium chloride solution in a case of diphtheria to reduce the risks of anaphylactic reaction due to the anti-diphtheria serum that he was ready to administer. To his great surprise, when the next day the laboratory results confirmed the diagnosis of diphtheria, the little girl was completely cured, before he could use the serum. Now what would be safer to administer to a child, a vaccine for diphtheria or a form of magnesium that comes from the sea? According to Dr. Delbet, magnesium chloride has a cytophilactic activity that no other magnesium salt has.

He credited the immuno-stimulant activity to the solution for this result, and he tested it in some other diphteric patients. All the patients were cured in a very short time (24-48 hours), with no after-effects. As magnesium chloride has no direct effect on bacteria (i.e. it is not an antibiotic), Neveu thought that its action was aspecific, immuno-enhancing, so it could be useful, in the same manner, also against viral diseases. So he began to treat some cases of poliomyelitis, and had the same wonderful results. He was very excited and tried to divulge the therapy, but he ran into a wall of hostility and obstructionism from "Official Medicine". Neither Neveu or Delbet (who was a member of the Academy of Medicine) was able to diffuse Neveu's extraordinary results. The opposition was total: Professors of Medicine, Medical Peer-Reviews, the Academy itself, all were against the two doctors.

It is little known, for instance, that magnesium increases the efficiency of white blood cells.

J. I. Rodale wrote in his book *Magnesium, The Nutrient that could change your life*, "Think what it could mean if we could induce the white cells in our blood to double their protective activity without any increase in numbers.[15] It would reduce sharply the possibility that invaders of the bloodstream could get by these defenders and do consequent damage to our systems. It would mean that the need for drugs to fight bacterial invasions would be just about eliminated. It would mean bringing the protective ability of everybody's blood up to the level that is now possessed by the superbly healthy individual."

Speaking further on white blood cells, J. I. Rodale continues, "There is, on the average, only one of these cells for each 150 of the red blood cells. These white corpuscles have a unique power. When the bloodstream is invaded by harmful bacteria or any other foreign matter, these white cells are somehow attracted to the source of the invasion, such as a wound, and go to work actually swallowing, and digesting the foreign matter and thus rendering it harmless. They do the same with any foreign bodies that infiltrate the bloodstream. They are the body's first and most important defense against all types of infection. But to increase the number of such cells circulating in the bloodstream would be a very dangerous thing. Leukemia, cancer of the blood, is marked by precisely such an increase.

"The destructive capacity of these cells is so great that their numbers must be kept at normal proportions for fear of the damage they might do our own systems if they got out of hand. Dr. Pierre Delbet claims that magnesium can do exactly this, strengthen the white cells without increasing their number.

"In a paper submitted in collaboration with Dr. Karalanopoulo to the French Academy of Science, September 6, 1915, titled *Cytophylaxis*, which means work done by the white cells or phagocytes in destroying invaders of the bloodstream. Delbet says, 'A solution of magnesium chloride at 12.1 parts per 1,000 gave extraordinary results. It increased the proportion of phagocytosis [killing microbes] by 75% as compared with the solution of sodium chloride at 8 parts per 1,000 which itself gave 63% more than the Locke-Ringer's solution. The increase is based on the number of polynucleates [white cells] as well as the phagocytic [germ-destroying] power of each of them.'

"These experiments prove that a solution of desiccated chloride of magnesium at 12.1 parts per 1,000 has a special effect on the white corpuscles, which is not the case with either physiological serum [a solution of chloride of sodium at seven parts per 1,000] or seawater, or the solution of Locke-Ringer which was considered best for maintaining the activity of cells.

"Consequently, a solution of chloride of magnesium was better than all the solutions previously used in the washing and dressing of wounds."

We invite you to further consider magnesium as the real "missing link," that is, to our physical health, because when our body is receiving this vital nutrient in proper amounts, the white cells perform exactly as J.I Rodale described.

Dr. Pierre Delbet used to give intravenous magnesium chloride solutions routinely to his patients with infections and for several days before any planned surgery and was surprised by many of these patients experiencing euphoria and bursts of energy.

This is not a fantasy but the celebrated work of Dr. Pierre Delbet,[16] called *Politique Préventif du Cancer.* Dr Raul Vergini, in Italy, says that epidemiological studies confirmed Delbet's views and demonstrated that the regions with soil more rich in magnesium had less cancer incidence.

He also says that in experimental studies the magnesium chloride solution was able to slow down the course of cancer in laboratory animals. Dr. Delbet demonstrated many years ago that magnesium chloride solution was a good therapy for a long list of diseases because it had a beneficial effect on the whole organism.

He obtained good results in: colitis (oral magnesium is contraindicated in severe intestinal disorders), angiocholitis and cholecystitis in the digestive apparatus; Parkinson´s disease, senile tremors and muscular cramps in the nervous system; acne, eczema, psoriasis, warts, itch of various origins and chilblains in the skin.

There was a strengthening of hair and nails, a good effect on diseases typical of the aged (impotency, prostatic hypertrophy, cerebral and circulatory troubles) and on diseases of allergic origin (hay-fever, asthma, urticaria and anaphylactic reactions).

*The clinical facts have, for the most part, been observed
by chance. My followers take much magnesium chloride.
They are enthusiasts propagandizing for it. Others adopt it,
partly, perhaps, because it often produces systemic excitation.
Among those who take it for its tonic action, several are
afflicted with various ailments which disappear, and they
report from time to time successes I did not expect.*

Dr. Pierre Delbet

Magnesium chloride, when used correctly, is the best weapon we have to defend the body, not only from infectious diseases of both viral and bacterial origin, but also from the deluge of toxic chemicals that invade our body everyday. Between its power to stimulate white blood cells and glutathione, we have a heavyweight non toxic *medicinal nutrient* that we can use without a prescription, because it is actually an essential *food* that the human body requires to function properly. Withholding magnesium from the diet is likened in importance to keeping oxygen out of the air that we breathe, and developing pharmaceutical substitutes to "improve" our respiration.

**Table 2: Recommended Dietary Allowances for
magnesium for children and adults.[17]
(National Institute of Health)**

Age (years)	Male (mg/day)	Female (mg/day)	Pregnancy (mg/day)	Lactation (mg/day)
1-3	80	80	N/A	N/A
4-8	130	130	N/A	N/A
9-13	240	240	N/A	N/A
14-18	410	360	400	360
19-30	400	310	350	310
31+	420	320	360	320

There is insufficient information on magnesium to establish a RDA for infants. For infants 0 to 12 months, the DRI is in the form of an Adequate Intake (AI), which is the mean intake of magnesium in healthy, breast fed infants. Table 3 lists the AI's for infants in milligrams (mg).

**Table 3: Recommended Adequate Intake for magnesium
for infants (National Institute of Health)**

Age (months)	Males and Females (mg/day)
0 to 6	30
7 to 12	75

Try taking the magnesium questionnaire in the Appendix section to see whether you are magnesium deficient. The time scale of mineral loss is long, it can be many months or even years before lack of exposure to certain elements is noted with respect to a person's state of health. Improving your mineral status by eating a whole food diet used to ensure that you get the right minerals in the correct quantities and proportions. But today supplementing is necessary because of increased chemical body burdens and the decreased mineral food values. Re-mineralization *will* improve the levels of long-term imbalances with truly life-changing results, and very quickly if administered in highly bioavaliable forms.

Magnesium rapidly distributes throughout the body following absorption. Normal plasma levels of magnesium range from about 1.6 to 2.1 nM.

In spite of its low cost or perhaps as a result of its low cost, magnesium is not given routinely to heart attack victims.[18] The "Myers' cocktail," which was used effectively for acute asthma attacks, migraines, fatigue (including chronic fatigue syndrome), fibromyalgia, acute muscle spasm, upper respiratory tract infections, chronic sinusitis, seasonal allergic rhinitis, cardiovascular disease, and other disorders, consisted of magnesium, calcium, B vitamins, and vitamin C. This treatment was made famous by Dr. Linus Pauling and many doctors around the world practice Orthomolecular Medicine with good clinical results. Dr. Pauling was well ahead of his time but today we need safer methods that are more practical and universal and easily administered to children.

Transdermal treatments are applicable and effective for almost all medical conditions and situations and can be as quick acting as an intravenous (IV) drip.

References

1   Source: Massachusetts Institute of Technology (www.sciencedaily.com/releases/2004/12/041219164941.htm)

2   King D, Mainous A 3rd, Geesey M, Woolson R. *Dietary magnesium and C-reactive protein levels.* J Am Coll Nutr. 2005 Jun 24(3):166-71.

3   *Is Agrobusiness Making Food Less Nutritious?* www.motherearthnews.com/library/2004_June_July/Is_Agribusiness_Making_Food_Less_Nutritious_

4   See: www.eatwild.com

5   Paul Mason. *Violence Prevention through Magnesium-Rich Water.* Healthy Water Association. www.mgwater.com/cyalettr.shtml

6   Altura BM, Introduction: importance of Mg in physiology and medicine and the need for íon selective electrodes. *Scand J Cliin Lab Invest Suppl*, vol. 217, pp. 5-9, 1994

7   Institute of Medicine, *Dietary Reference Intake for Calcium, Phosphorus, Magnesium, vitamin D and Flouride*, National Academy Press, Washington DC, 1997

8   Acid-Base Status Affects Renal Magnesium Losses in Healthy, Elderly Persons. **Ragnar Rylander, Thomas Remer, Shoma Berkemeyer and Jürgen Vormann.** 2006 *J. Nutr.* 136:2374-2377, September 2006

9   U.S. Dept. of Health, Agency for Toxic Substances and Disease Registry, Division of Toxicology, December 16, 1991. www.mgwater.com/fluoride.shtml

10   www.lef.org

11   Oral Magnesium Chloride, Magnesium Citrate Magnesium Gluceptate, Magnesium Gluconate, Magnesium Hydroxide, Magnesium Lactate, Magnesium Oxide, Magnesium Pidolate, Magnesium Sulfate.

12   chelate: a compound having a ring structure that usually contains a metal ion held by coordinate bonds

13   http://ods.od.nih.gov/factsheets/magnesium.asp#en9#en9

14   Crippa G, Sverzellati E, Giorgi-Pierfranceschi M, et al. *Magnesium and cardiovascular drugs: interactions and therapeutic role.* Ann Ital Med Int. 1999 Jan; 14(1):40-5.

15   Experimentally Magnesium has been shown to have a role in myocardial salvage, possibly by inhibiting calcium influx to ischaemic myocytes and/or by reducing coronary tone. It has also been shown to increase the threshold for depolarisation of cardiac myocytes, theoretically reducing the risk of malignant arrhythmia. In healthy humans it can reduce peripheral vascular resistance and increase cardiac output with no effect on cardiac work.

16   In another communication to the French Academy of Medicine (September 7, 1915), Dr. Delbet describes research that proved the effectiveness of magnesium within

the body. He injected 150 cc. of a solution of magnesium chloride into the vein of a dog, taking a blood sample before the injection and a second one 35 minutes afterward. Then the white corpuscles were presented with microbes from the same culture to see their effect on or power to destroy them. Five hundred white cells in the first sample destroyed 245 microbes. Five hundred white cells from the second destroyed 681. This increase in microbe-killing under the influence of magnesium chloride was 180% over the other solutions. More experiments were performed; in one there was an increase to 129%, in another, 333%.

17   Back in 1915, a French surgeon, Prof.Pierre Delbet, M.D., was looking for a solution to cleanse wounds, because he had discovered that the traditional antiseptic solutions actually mortified tissues and facilitated the infection instead of preventing it. He tested several mineral solutions and discovered that a magnesium chloride solution was not only harmless for tissues, but it had also a great effect over leucocytic activity and phagocytosis; so it was perfect for external wounds treatment. Prof. Delbet wrote two books, Politique Preventive du Cancer (1944) and L'Agriculture et la Santé (1945), in which he stated his ideas about cancer prevention and a better living. The first is a well documented report of all his studies on magnesium chloride.

18   See: http://ods.od.nih.gov/factsheets/magnesium.asp#en4#en4

19   According to Dr. Seelig, "In Finland, which has a very high death rate from IHD, there is a clear relationship with heart disease and the amount of magnesium in the soil (Karppanen and Neuvonen, 1973). In eastern and in northern Finland, where the soil content is about a third of that found in southwestern Finland (Karppanen et al., 1978) the mortality from ischemic heart disease is twice as high as is that in the southwest. Ho and Khun (1976/1980) surveyed factors that might be contributory both to the rising incidence of cardiovascular disease in Europe, and the falling levels of magnesium both in the soil and in the food supply. They commented that in Finland, which has the highest cardiovascular death rate in Europe, the dietary supply of magnesium had decreased by 1963 to a third of the intake common in 1911 (H. Katz, 1973). In contrast, in Japan with its low cardiac death rate, the daily magnesium intake was cited as 560 mg (Holtmeier, 1969a, 1973)." Recent evidence suggests that vitamin E enhances glutathione levels and may play a protective role in magnesium deficiency-induced cardiac lesions.

4

||||||||||||||||||||||||||||||||||||||||||||||||||||||||||||||||||||||||||||||

Magnesium and Calcium

*Calcium and magnesium are opposites in their effects
on our body structure. As a general rule, the more rigid and
inflexible our body structure is, the less calcium and the more
magnesium we need.*

D
r. Garry Gordon wrote, "If you have compromised cell membranes or low ATP production for any reason, then the cell has trouble maintaining the normal gradient. This is because the usual gradient is 10,000 times more calcium outside of cells than inside; when this is compromised you will have increased intracellular calcium, which seems to always happen at the time of death. Whenever intracellular calcium is elevated, you have a relative deficiency of magnesium, so whenever anyone is seriously ill, acute or chronic, part of your plan *must* be to restore magnesium."

*The ratio of calcium to magnesium is vital for
cell membranes and the Blood Brain Barrier.*

Countries with the highest calcium to magnesium ratios (high calcium and low magnesium levels) in soil and water have the highest incidence of cardiovascular disease. At the top of the list is Australia. In contrast, in Japan with its low cardiac death rate, the daily magnesium intake was cited as high as 560 milligrams. The human populations that consume the most calcium tend to have the highest mortality rates in the world. The Scandinavian countries, the USA and New Zealand are the dairy consuming countries and mortality rates soar in these countries. In Japan where the consumption of calcium from dairy products is the lowest on the planet so are the mortality rates. "There is a lot of calcium in most diets, and even a relatively small amount of calcium supplementation, taken on a regular basis, can result in undesirable, rocklike, non-biologic deposits of calcium in the tissues," says Dr. Thomas E. Levy.

*The widespread shortage of magnesium, not
calcium, in the western diet is attributed to the
high rates of sudden-death heart attack.*

Adequate levels of magnesium are essential for the heart muscle. Those who die from heart attacks have very low magnesium but high calcium levels in their heart muscles. Patients with coronary heart disease who have been treated with large amounts of magnesium survived better than those with other drug treatments. Magnesium dilates the arteries of the heart and lowers cholesterol and fat levels.

Magnesium taken in proper dosages can solve
the problem of calcium deficiency.

Dr. Nan Kathryn Fuchs

Author of *The Nutrition Detective*

It is magnesium that controls the fate of potassium and calcium in the body. If magnesium is insufficient, potassium and calcium will be lost in the urine and calcium will be deposited in the soft tissues (kidneys, arteries, joints, brain, etc.). In properly balanced proportions, magnesium and calcium have *complementary* effects on many of the body's chemical pathways. *Both* are required for optimum body function and health.

Calcium causes muscles to contract, while
magnesium helps them relax.

Magnesium and calcium are paired minerals. Several studies have reported that increasing calcium in the diet significantly *reduces* the absorption of magnesium. Calcium intakes above 2.6 grams per day may reduce the uptake and utilization of magnesium by the body thus increasing magnesium requirements. So much stress is placed on the importance of calcium by the dairy industry that we have, in fact, compromised magnesium absorption.

Up to 30% of the energy of cells is used to
pump calcium out of the cells.

A *healthy* cell has high magnesium and low calcium levels. The higher the calcium level and the lower the magnesium level in the extra-cellular fluid, the harder is it for cells to pump the calcium out. The result is that with low magnesium levels the mitochondria gradually calcify and energy production decreases. Our biochemical age could theoretically be determined by the ratio of magnesium to calcium within our cells.

Magnesium is the mineral of rejuvenation and prevents the calcification of our organs and tissues that is characteristic of the old-age related degeneration of our body.

Without sufficient magnesium, calcium can collect in the soft tissues and cause arthritis. Not only does calcium collect in the soft tissues of arthritics, it is poorly, if at all, absorbed into their blood and bones. Some researchers estimate that the American ratio of calcium to magnesium is actually approaching 6:1, while the recommendation for healthy living is actually 2:1. But even 2 parts of calcium to 1 part of magnesium is probably too high, since current research on the Paleolithic or caveman diets show that the ratio they used to eat was 1:1.[1]

A diet high in dairy and low in whole grains can lead to excess calcium in the tissues and a magnesium deficiency.[2]
Dr. Nan Kathryn Fuch

According to Dr. P. Kaye, Emergency Department, Bristol Royal Infirmary, UK, "Magnesium acts as a smooth muscle relaxant by altering extracellular calcium influx and intracellular phosphorylation reactions. It may also attenuate the neutrophilic burst associated with inflammatory bronchoconstriction by attenuating mast cell degranulation. The principal trigger for this degranulation is a rise in intracellular calcium, which is antagonized by magnesium. It has been shown experimentally to augment the bronchodilatory effect of salbutamol and to inhibit histamine induced bronchospasm. Magnesium should be used as a safe, easy-to-administer and effective second line agent in acute severe asthma."[3]

Calcium can accumulate in heart valves (mitral valve), and can become a concretion in the kidneys and become a stone, a condition that affects 1 in 12 Americans.

Magnesium deficiency factor in osteoporosis and tooth decay

Western medical authorities claim that the widespread incidence of osteoporosis and tooth decay in western countries can be prevented with a high calcium intake. Yet, it is evident that this approach will only exacerbate an already alarming trend. Asian and African populations with a *low* daily intake of calcium (about 300 mg), experience very little osteoporosis. Bantu women with an intake of 200 to 300 mg of calcium daily have the lowest incidence of osteoporosis in the world.[4]

In western countries with a high intake of dairy products the average calcium intake is about 1000 mg. With a low magnesium intake, calcium moves out of the bones to increase tissue levels, while a high magnesium intake causes calcium to move from the tissues into the bones. Thus high magnesium levels leads to bone mineralization.

Dr. Karen Kubena, associate professor of nutrition at Texas A & M University, indicates that even if you monitor your magnesium level like a maniac, you're still at risk for migraines if your calcium level is out of whack. It seems that higher than normal blood levels of calcium cause the body to excrete the excess calcium, which in turn triggers a loss of magnesium. "Let's say you have just enough magnesium and too much calcium in your blood. If calcium is excreted, the magnesium goes with it. All of a sudden, you could be low in magnesium," says Dr. Kubena.[5]

If not enough magnesium is taken with calcium, it will cause more harm than good. The unabsorbed calcium can lodge anywhere in the body and provoke practically any disease. For instance, if it lodges in your bones and joints, it leads to some forms of arthritis; if it lodges in you heart, it leads to arterial lesions; it provokes respiratory problems if it lodges in your lungs, etc.

Despite the crucial relationship between calcium and magnesium, a recently published study announced that most U.S. children don't get enough *calcium* in their diets, and pediatricians should intervene to help remedy the problem. These guidelines were issued in Feb. 2006 by the American Academy of Pediatrics.[6]

"The proportion of children, who receive the recommended amounts of calcium declines dramatically after the second year of life, reaching a nadir during adolescence," said Dr. Nancy F. Krebs, of the University of Colorado Health Sciences Center, Department of Pediatrics, located in Denver, who headed the academy committee that wrote the guidelines.

Adolescent girls are faring the worst, Dr. Krebs and colleagues reported. Only about 10% of girls ages 12 to 19 are getting the recommended amount of calcium. For boys, the figure is about 30%. However, not a word is mentioned about magnesium as the committee goes on to recommend *increasing* calcium intake through the use of fortified foods and calcium supplements. Is a medical crime being committed when these pediatricians fail to address the crucial relationship between magnesium and calcium? Our affirmative answer is sustained when reviewing the materials presented below.

Experts say excessive calcium intake may be unwise in light of recent studies showing that high amounts of the mineral may increase risk of prostate cancer. "There is reasonable evidence to suggest that calcium may play an important role in the development of prostate cancer," says Dr. Carmen Rodriguez, senior epidemiologist in the epidemiology and surveillance research department of the American Cancer Society (ACS).

Rodriguez says that a 1998 Harvard School of Public Health study of 47,781 men found those consuming between 1,500 and 1,999 mg of calcium per day had about double the risk of being diagnosed with metastatic (cancer that has spread to other parts of the body) prostate cancer as those getting 500 mg per day or less. And those taking in 2,000 mg or more had over *four times the risk* of developing metastatic prostate cancer as those taking in less than 500 mg.

The recommended daily allowance (RDA) of calcium is
1,000 mg per day for men, and 1,500 mg for women.

Later in 1998, Harvard researchers published a study of dairy product intake among 526 men diagnosed with prostate cancer and 536 similar men not diagnosed with the disease. That study found a 50% increase in prostate cancer risk and a near doubling of risk of metastatic prostate cancer among men consuming high amounts of dairy products, likely due, say the researchers, to the high total amount of calcium in such a diet. The most recent Harvard study on the topic, published in October 2001, looked at dairy product intake among 20,885 men and found men consuming the most dairy products had about 32% higher risk of developing prostate cancer than those consuming the least.

According to the University of Florida Shands Cancer Center, a high level of calcium in the blood, called hypercalcemia,[7] may become a medical emergency. This disorder is most commonly caused by cancer or parathyroid disease but underneath the primary etiology is probably magnesium deficiency.

Hypercalcemia is commonly attributed to either the cancer treatment or the cancer itself and may make it difficult for doctors to detect hypercalcemia when it first occurs. This disorder can be severe and difficult to manage especially because doctors have not a clue about the underlying relationship between excess calcium and low levels of magnesium. Severe hypercalcemia is a medical emergency that can be avoided if magnesium levels are brought up to normal.

Calcium competes with zinc, manganese, magnesium, copper and iron for absorption in the intestine and a high intake of one can reduce absorption of the others.

Because of the totally distorted way medical science relates to magnesium the medical profession makes mistakes with calcium. It's still common to hear the assumption about calcium's ability to help prevent osteoporosis (weakening of the bones usually associated with aging). *The fact is that it's the increasing of magnesium intake that improves bone density*[8] in the elderly and reduces the risk of osteoporosis. "Higher Magnesium intake through diet and supplements was positively associated with total-body bone mineral density (BMD) in older white men and women. For every 100 mg per day increase in magnesium, there was an approximate 2% increase in whole-body BMD,"[9] said Dr. Kathryn Ryder.

Magnesium is essential for proper calcium absorption and is an important mineral in the bone matrix.

"Bones average about 1% phosphate of magnesium and teeth about 1% phosphate of magnesium. Elephant tusks contain 2% of phosphate of magnesium and billiard balls made from these are almost indestructible. The teeth of carnivorous animals contain nearly 5% phosphate of magnesium and thus they are able to crush and grind the bones of their prey without difficulty," wrote Otto Carque (1933) in *Vital Facts About Foods.*

Some people, like a spokesperson for the UK-based charity, the National Osteoporosis Society, continue to think that "magnesium deficiency is, in fact, very rare in humans." So they cannot get it through their neural circuits that magnesium deficiency, and not calcium deficiency plays a key role in osteoporosis.

Thus it is no surprise when we find more studies suggesting that high calcium intake had no preventive effect on alteration of bone metabolism in magnesium deficient rats[10] and that not only severe but also moderate dietary restriction of magnesium results in qualitative changes in bones in rats.[11]

The results from some of these studies may be surprising to some. While we have no reason to question the importance of calcium in bone strength, we have plenty of reason to doubt the value of consuming large amounts of calcium that are currently being recommended for adults and young people alike.[12]

One of the most important aspects of the disease
osteoporosis has been almost totally overlooked.
That aspect is the role played by magnesium.

Dr. Lewis B. Barnett

While most sources understand that calcium is important in the growth and development of children, little attention is paid to the role of magnesium or magnesium deficiency or the need to maintain the intricate balances of each (and other nutrients as well).

Back in the 1950's Dr. Barnett examined the bone content of healthy people and compared it with the content of people suffering from severe osteoporosis. He found there was little difference among the calcium, phosphorus, and fluoride content of the bones of the individuals. The magnesium content in the bones of the healthy people, however, was 1.26 %. That of the osteoporosis victims was slightly less than half that amount, at .62 %.

Many years ago Dr. Barnett conducted tests on 5,000 people and found about 60% of them deficient in magnesium. Today we find MIT placing that number officially at 68%. How is it that so many in the medical profession are oblivious to this clinical reality and go on acting as though magnesium deficiency in the general population is rare?

Magnesium status is important for regulation of calcium
balance through parathyroid hormone-mediated reactions.[13]

The current focus on increased need for calcium in a chronically magnesium deficient population, can easily push those already receiving adequate amounts of calcium in their daily diets over the edge to reaching too high levels, thus causing depletion of magnesium with its many residual side-effects.

Yet, the American Diabetes Association in their 2006 guidelines for diabetes and pre-diabetes, when making treatment and nutritional recommendations, joined pediatricians in maintaining the general medical status quo, by not calling attention to this very correctable health concern and recommending that magnesium be addressed in a significant way. This is despite the increasing evidence over the years that magnesium is even more deficient in diabetics and that current dietary recommendations do not address the issue.

This medical review is important exactly because, as of late 2006, large segments of the medical establishment are yet overlooking the relationship of magnesium and calcium and, intentionally or otherwise, are leading the public to the Church of Iatrogenic Disease, a place where billions of dollars are made at the expense of *natural* and otherwise sustainable health.

Let us pray.

Despite the fact that serum levels are not the best indicator of adequate magnesium presence in the body, some studies have shown that when magnesium deficiency was induced in humans, the earliest sign was decreased serum magnesium levels (hypomagnesemia). Over time, serum calcium levels also began to decrease (hypocalcemia) despite adequate dietary calcium.

Hypocalcemia persisted despite increased parathyroid hormone (PTH) secretion. Usually, increased PTH secretion quickly results in the mobilization of calcium from bone and normalization of blood calcium levels. As the magnesium depletion progressed, PTH secretion diminished to low levels. Along with hypomagnesemia, signs of severe magnesium deficiency included hypocalcemia, low serum potassium levels (hypokalemia), retention of sodium, low circulating levels of PTH, neurological and muscular symptoms (tremor, muscle spasms, tetany), loss of appetite, nausea, vomiting, and personality changes.[14] Hypercalcemia can cause magnesium deficiency and wasting.[15]

Magnesium is required for the activation of alkaline phosphatase, an enzyme involved in forming calcium crystals in bone, and for the conversion of vitamin D into its biologically active form.

It is medical wisdom that tells us that magnesium is actually the key to the body's proper assimilation and use of calcium, as well as other important nutrients. If we consume too much calcium, without sufficient magnesium, the excess calcium is not utilized correctly and may actually become toxic, causing painful conditions in the body. Hypocalcemia is a prominent manifestation of magnesium deficiency in humans (Rude et al., 1976). Even mild degrees of magnesium depletion significantly decreases the serum calcium concentration (Fatemi et al., 1991).

The adverse effects of excessive calcium intake may include high blood calcium levels, kidney stone formation and kidney complications.[16] Elevated calcium levels are also associated with arthritic/joint and vascular degeneration, calcification of soft tissue, hypertension and stroke, and increase in VLDL triglycerides, gastrointestinal disturbances, mood and depressive disorders, chronic fatigue, and general mineral imbalances including magnesium, zinc, iron and phosphorus. High calcium levels interfere with vitamin D and subsequently inhibit the vitamin's cancer protective effect unless extra amounts of vitamin D are supplemented.[17]

William R. Quesnell, author of *Minerals: The Essential Link to Health*, said, "Most people have come to believe nutrition is divisible, and that a single substance will maintain vibrant health. The touting of calcium for the degenerative disease osteoporosis provides an excellent example. Every day the media, acting as proxy for the milk lobby, sells calcium as a magic bullet. Has it worked? Definitely for sales of milk, but for the State of Health in America, it has been a disaster. When you load up your system with excess calcium, you shut down magnesium's ability to activate thyrocalcitonin, a hormone that under normal circumstances would send calcium to your bones."

References

1   Eades M, Eades A, *The Protein Power Lifeplan*, Warner Books, New York, 1999

2   The source of menstrual cramps may come from eating too much cheese, yogurt, ice cream or milk, combined with insufficient whole grains and beans. Or it could come from taking too much calcium without enough magnesium. Modifying diet and increasing magnesium supplementation may allow menstrual cramps to disappear.

3   Kaye, P. O'Sullivan, I. The role of magnesium in the emergency department. Emergency Department, Bristol Royal Infirmary, Bristol, UK Emerg Med J 2002; 19:288-291

4   See: http://list.weim.net/pipermail/holisticweim/2001-July/001023.html

5   See: www.mgwater.com/prev1801.shtml

6   Article: *Pediatricians Say That Most US Kids Don't Get Enough Calcium* http://www.medpagetoday.com/Pediatrics/GeneralPediatrics/dh/2624

7   Signs and symptoms of hypercalcemia may include: Nausea, Fatigue, Vomiting, Lethargy, Stomach pain, Moodiness, Constipation, Irritability, Anorexia, Confusion, Excessive thirst, Extreme muscle weakness, Dry mouth or thoat, Irregular heart beat, Frequent urination, Coma

8   Stendig-Lindberg G. Tepper R. Leichter I. Trabecular bone density in a two year controlled trial of peroral magnesium in osteoporosis. Department of Physiology and Pharmacology, Sackler Faculty of Medicine, Tel Aviv University, Israel. *Magnes Res.* 1993 Jun;6(2):155-63.

9   *Journal of the American Geriatric Society* (November, Vol 53, No 11, pp 1875-1880).

10   We examined the effects of high calcium (Ca) intake on bone metabolism in magnesium (Mg)-deficient rats. Male Wistar rats were divided into three groups, with each group having a similar mean body weight, and fed a control diet (control group), a Mg-deficient diet (Mg-deficient group) or a Mg-deficient Ca-supplemented diet (Mg-deficient Ca-supplemented group) for 14 d. Femoral Ca content was significantly lower in the Mg-deficient Ca-supplemented group than in the control group and Mg-deficient group. Femoral Mg content was significantly lower in the Mg-deficient group

and Mg-deficient Ca-supplemented group than in the control group. Furthermore, femoral Mg content was significantly lower in the Mg-deficient Ca-supplemented group than in the Mg-deficient group. Serum osteocalcin levels (a biochemical marker of bone formation) were significantly lower in the two Mg-deficient groups than in the control group. As a biochemical marker of bone resorption, urinary deoxypyridinoline excretion was significantly higher in the Mg-deficient Ca-supplemented group than in the control group and Mg-deficient group. The results in the present study suggest that high Ca intake had no preventive effect on alteration of bone metabolism in Mg-deficient rats. Effects of high calcium intake on bone metabolism in magnesium-deficient rats. Magnes Res. 2005 Jun;18(2):97-102.

11   Br J Nutr. The effect of moderately and severely restricted dietary magnesium intakes on bone composition and bone metabolism in the rat.1999 Jul;82(1):63-71.

12   In particular, these studies suggest that high calcium intake doesn't actually appear to lower a person's risk for osteoporosis. For example, in the large Harvard studies of male health professionals and female nurses, individuals who drank one glass of milk (or less) per week were at no greater risk of breaking a hip or forearm than were those who drank two or more glasses per week. Other studies have found similar results. Additional evidence also supports the idea that American adults may not need as much calcium as is currently recommended. For example, in countries such as India, Japan, and Peru where average daily calcium intake is as low as 300 mg/day (less than a third of the US recommendation for adults, ages 19-50), the incidence of bone fractures is quite low. Of course, these countries differ in other important bone-health factors as well - such as level of physical activity and amount of sunlight - which could account for their low fracture rates. *Calcium in Milk*, Harvard School of Public Health; For more, see: www.hsph.harvard.edu/nutritionsource/calcium.html

13   Northwestern University; Nutrition Fact Sheet: www.feinberg.northwestern. edu/nutrition/factsheets/magnesium.html

14   Shils ME. Magnesium. In: Shils M, Olson JA, Shike M, Ross AC, eds. Nutrition in Health and Disease. 9th ed. Baltimore: Williams & Wilkins; 1999:169-192.

15   Other causes of renal magnesium wasting include aldosterone excess, most likely through chronic volume expansion, causing increased magnesium excretion; hypercalcemia due to increased competition for reabsorption with magnesium; Hypercalcemia inhibits magnesium reabsorption, probably through competition for passive transport through the renal system. Hypomagnesemia; Mahendra Agraharkar, MD,FACP Updated: June 20, 2002 http://www.emedicine.com/med/topic3382.htm

16   New York State Department of Health; www.health.state.ny.us/diseases/ conditions/osteoporosis/qanda.htm

17   Accu-Cell Nutrition; Calcium and Magnesium, www.acu-cell.com/acn.html

5

||||||||||||||||||||||||||||||||||||||||||||||||||||||||||||||||||||||||||

Disease, and Preventive Care

M any people find it tempting to oversimplify disease. Dr. Hulda Clark, for example, wrote the following in the beginning of her book, *The Cure for All Diseases*:

"No matter how long and confusing is the list of symptoms a person has, from chronic fatigue to infertility to mental problems, I am sure to find only two things wrong: they have in them pollutants and/or parasites. I never find lack of exercise, vitamin deficiencies, hormone levels or anything else to be a primary causative factor. So the solution to good health is obvious:

Problem	**Simplest Cure**
Parasites	Electronic and herbal treatment
Pollution	Avoidance

I personally can appreciate this kind of oversimplification because there was a time when my wife and I used to laugh at my own oversimplification: standing on one foot and a closed heart. I wrote my book *HeartHealth* about the situation with our hearts (that medicine pays no attention to). On my Biogenic Medicine site I have a section with pictures about the structural damage of defying gravity with the common posture of shifting our body weight to one side. Next time you go into a bank just watch everyone and see for yourself. Few today stand evenly on two feet and chiropractors do a booming business for this and of course, other reasons.

Magnesium deficiency certainly qualifies as a principal cause of disease. No matter what we do with our hearts, postures, or medical treatments, there is simply nothing we can do to adequately *enhance* our State of Health when magnesium supplies are less than adequate in our bodies. Only now is it dawning on us just how important magnesium is.

The art and science of helping people through medicine is a daunting task that involves the knowledge, experience and understanding to encompass a whole range of factors.

The health status of the digestive system and the
kidneys significantly influence magnesium status.

Without adequate levels of magnesium, our hearts definitely suffer. Magnesium coordinates the activity of the heart muscle as well as the functioning of the nerves that initiate the heartbeat.

Ponder that last thought for a moment. It bears repeating.

Magnesium coordinates the activity of the
heart muscle as well as the functioning of the
nerves that initiate the heartbeat.

It also helps keep coronary arteries from spasming, an action that can cause the intense chest pain known as angina. The most alarming trend in the past century has been the sharp increase in sudden deaths from ischemic heart disease (IHD), particularly in middle-aged men, and the increasing number of younger men who suddenly develop myocardial infarctions, cardiac arrhythmias, or arrests.

Dr. Mildred Seelig, of New York University Medical Center, wrote twenty-five years ago in her book, *Magnesium Deficiency in the Pathogenesis of Disease*, that magnesium deficiency was probably the common etiologic (i.e., causal) factor in the increased incidence of sudden infant deaths, infantile myocardial infarction and arteriosclerosis, and the same diseases that become manifest later in life. It is also suggested that magnesium deficiency might cause or predispose to certain skeletal and renal diseases, all of which can coexist.

It has been said that many sudden deaths following vaccination can be prevented by sufficient vitamin C and vitamin A reserves (or supplementation), but we should probably take a good look at magnesium deficiency as a crucial factor in sudden death.

In adults it has been seen how high levels of mercury in the heart becomes dangerous. Since deficient magnesium and high levels of mercury are related, childhood vaccines containing thimerosal would be especially dangerous to administer in children deficient in magnesium.

According to Dr. Seelig, "In Finland, which has a very high death rate from heart disease, there is a clear relationship between heart disease and the amount of magnesium in the soil (Karppanen and Neuvonen, 1973). In eastern and in northern Finland, where the soil content is about a third of that found in southwestern Finland (Karppanen *et al.*, 1978) the mortality from ischemic heart disease is twice as high as is that in the southwest. Finland, which has the highest cardiovascular death rate in Europe, has experienced a huge decrease in magnesium intake and today stands at less than a third of what it was in 1911 (H. Katz, 1973).

Preventive Medicine

Typically, if you are a mainstream allopathic physician, preventive medicine is limited to elective stress testing, mammography, screening for prostate-specific antigen, periodic lipid profiling and giving cursory attention to life-style changes and diet. This kind of medicine offers only a shadow of what preventive medicine needs to be.

According to the New York Times, "New evidence keeps emerging that the medical profession has sold its soul in exchange for what can only be described as bribes from the manufacturers of drugs and medical devices."[1] A soulless medicine's first crime is its failure to seriously embrace preventive medicine and avoid disease in the first place. None of the pharmaceutical companies are interested in preventive medicine for they make all their money from sickness and disease.

Preventive medicine is as important as any other type of medicine. In ancient China the oriental medical doctors were paid for keeping people well thus patients stopped paying when they fell sick. Allopathic medicine, through its major misunderstanding, denial and neglect of preventive medicine, is directly contributing to modern man becoming the most diseased population in the history of the world.

Doctors and health officials really have no idea that the vast majority of diseases can be prevented and even cured without drugs or surgery. In fact, the idea of natural, non-surgical treatments and cures of disease would most likely sound absurd to them. Their idea of preventive medicine starts and ends with pharmaceutical vaccines, which contain toxic chemicals like mercury that, instead of having a preventive or even healing effect, may actually be contributing greatly to the swift rise in chronic diseases.

Allopathic medicine could do much to redeem its soul if it would face all the evidence that has been building up which suggests that we have to add adequate intake of magnesium—the "forgotten electrolyte"—to our list of preventive health measures.

Ensuring adequate magnesium intake, through a combination of dietary sources, oral supplementation, and most importantly, the use of transdermal methods of application, constitutes a crucial step toward health that is justified by scientific and clinical data.

The American public consumes less magnesium than necessary for good health. As such, magnesium supplementation is indicated for almost everyone. It is the single greatest thing we can do to avoid serious diseases or recover from the ones we already have. The magnesium factor is not the only factor, but there is no other single element that can compare to its pervasiveness in areas of health maintenance.

The use of magnesium as a vital (food) nutrient, and as a preventive, clinical and emergency medicine would probably spell disaster for the $1.6 trillion medical industry in the United States.

Single-handedly, it could eliminate the need for hundreds of billions of dollars in medical expenditures and diminish a mountain of pain, misery and preventable death. When combined with some other medical essentials, such as vitamin C, proven antioxidants and minerals like selenium, zinc, organic foods, and full hydration with pure water, the reduction in medical costs would be staggering to the industry. On the other hand, the increase in the general health and well-being of the general population would have a profoundly beneficial effect on such factors as productivity, and the economy as a whole.

A mountain of evidence supports these statements. For instance, Dr. Frank D. Gilliland, professor of preventive medicine at the Keck School of Medicine at the University of Southern California, and his colleagues, monitored more than 2,500 pre-teens and teen-agers in a dozen Southern California communities, tracking children's food intake while measuring lung function (how well their lungs work). The team found that children who eat lots of antioxidant-rich fruit and juices — as well as those who get an abundance of magnesium and potassium — perform better on lung function tests than children who intake less of the nutrients.[2] vitamin C plays a big role in lung development, Gilliland says, while vitamins E and A also appear helpful, especially in children with asthma. This information is especially important for children who live in cities and other areas where air pollution is a problem.

According to Dr. Ronald Elin and Dr. Robert Rude, "Refining and processing of grains and other foodstuffs typically results in loss of 70% or more of the magnesium content (as well as other nutrients). The conversion of wheat into flour results in a loss of 82% of magnesium. Refining rice into polished rice sacrifices 83% of the magnesium. Milling corn into corn starch loses 98% of the magnesium. When soy beans are cooked, they lose 69% of their magnesium. Quick-cooking oatmeal provides only about 15% of the magnesium obtained from the slow-cooking cereal. As the nation's eating habits have gone from freshly prepared items made in the home to prepared, processed meals and "fast foods" taken on the run, the magnesium content of the food has plummeted."[3]

Our bodies simply cannot extract adequate
nutrition from today's "fast" and processed foods.

Dr. David Thomas, who researched government records, found that the levels of magnesium in the average rump steak have dropped 7%. Milk appears to have lost 2% of its calcium and 21% of its magnesium.[4] According to the analysis, cheddar provides 9% less calcium today, 38% less magnesium and 47% less iron, while parmesan shows the steepest drop in nutrients, with magnesium levels down by 70% and iron all gone compared with content in the years up to 1940. Though some of this information is being contested,[5] academics in the US and Denmark have also reported significant changes in the nutritional profile of modern foods.

Studies show that people who eat at least three servings of whole grains a day have a lower risk of heart disease, diabetes and cancer and seem to maintain a healthy weight more easily. Since refined grains, such as white flour, have their innermost and outermost layers (their germ and bran) removed, they are stripped of a great portion of their minerals. Whole grains are not only higher in fiber but contain four times the magnesium and zinc and twice the selenium.[6]

Only about 15 to 25% of children eat
the recommended amount of magnesium.

Even in individuals who are unwilling to make prudent changes in their diets and sedentary habits, the administration of certain nutrients and/or drugs may help to prevent or postpone the onset of Type II diabetes. The evident ability of fiber-rich cereal products to decrease diabetes risk, as documented in prospective epidemiological studies, is most likely mediated by the superior magnesium content of such foods. High-magnesium diets have preventive (though not curative) activity in certain rodent models of diabetes; conversely, magnesium depletion provokes insulin resistance.[7]

A non-drug abortive approach to migraine attacks has been the use of 1g magnesium sulfate through a slow intravenous push during an acute migraine with 85% effectiveness.[8]

Deficiencies in magnesium affect all people, leaving them vulnerable to developing acute and chronic conditions. Humans are genetically designed to be strong and healthy. Illness is not our natural state, but instead, is indicative of nutritional deficiencies, mineral imbalances, destructive *thinking*, or harmful toxicity levels that call for our attention and correction.

Habitual patterns of living and nutrition, which establish our standards of normalcy, can break down our natural state of healthiness. It is thought that each person is especially susceptible to certain diseases when compared to other people, but each person is also more resistant to certain diseases when compared to others. While each person could be thought to have certain weaknesses and strengths, nobody is strong enough to live without air, water, or *magnesium* for very long.

Though in the long run food can be considered one of the best medicines, it is difficult to recover fully from magnesium deficiencies through changes in diet only. Even the use of oral magnesium supplementation is slow and needs to be augmented with quick-acting transdermal methods of application and by intramuscular and intravenous magnesium in emergency situations.

A magnesium deficiency is closely associated with cardiovascular disease.[9] Lower magnesium concentrations have been found in heart attack patients[10] and administration of magnesium[11] has proven beneficial in treating ventricular arrhythmias.[12,13,14,15] Fatal heart attacks are more common in areas where the water supply is deficient in magnesium and the average intake through the diet is often significantly less than the 200-400 milligrams required daily.[16]

Magnesium is proving to be very important in the maintenance of heart health and in the treatment of heart disease. Magnesium, calcium, and potassium are all effective in lowering blood pressure.[17,18,19,20] Magnesium is useful in preventing death from heart attack and protects against further heart attacks.[21,22] It also reduces the frequency and severity of ventricular arrhythmias and helps prevent complications after bypass surgery.

Prenatal magnesium factors

Using magnesium as a preventive medicine starts in pregnancy for there are significant benefits of magnesium for preeclampsia and eclampsia. Eclampsia and preeclampsia are the leading causes of death for pregnant women and their fetuses, particularly in developing countries. Physicians believe the high blood pressure, swelling, and protein in the urine associated with preeclampsia lead to the convulsions and coma of eclampsia. Magnesium is most effective at preventing eclamptic seizures. Now magnesium sulfate is being used increasingly to treat preeclampsia as well, with the hope it will prevent eclampsia. A study published in the June 1, 2002, issue of *The Lancet* confirms this hope.[23]

*Women receiving magnesium sulfate had a 27%
lower risk of premature detachment of the placenta.*

Drs. Elin and Rude sum up the universal need for all health care practitioners to become more aware of the crucial role of magnesium and the need to start practicing preventive medicine with magnesium. "The common denominator between magnesium intake and these diseases (see below list), as well as diabetes mellitus, migraine, and premenstrual syndrome is chronic latent magnesium deficiency. It is reasonable to suspect chronic latent magnesium deficiency in any patient with cardiovascular disease, diabetes mellitus, migraine, osteoporosis, or premenstrual syndrome, or risk factors for any of these conditions."[24]

*As shown in the table on the next page, important nutrients
disappear when whole wheat or other grains are refined. As
this baker's dozen shows, the losses can be dramatic. For
example, refined wheat flour has only 5% of the vitamin E as
whole wheat flour.*

It is reasonable as well to suspect chronic latent magnesium deficiency in anyone eating white bread and white rice and other heavily processed foods.

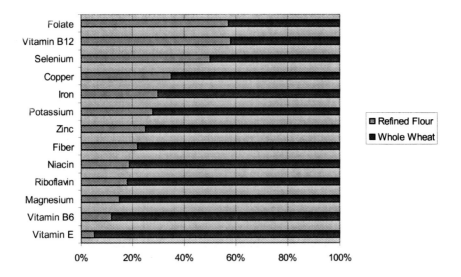

Harvard University

People who regularly eat whole grains develop cancer less often than those who don't. A 1998 overview of 40 studies that looked at 20 types of cancer linked consumption of whole grains with reduced risks of stomach, colon, mouth, gallbladder, and ovarian cancers.[25]

A close study of all the literature yields a long list of clinical situations associated with magnesium deficiency. Magnesium intake decreases significantly in persons age 70 and older — precisely those at highest risk for many of the diseases associated with chronic latent magnesium deficiency. Suboptimal intake of magnesium has been associated with the following prevalent and potentially serious conditions.

ADD/ADHD	Diabetes- Type I and II
Alzheimer's	Eating disorders- Bulimia, Anorexia
Angina	Fibromyalgia
Anxiety disorders	Gut disorders- including peptic ulcer, Crohn's disease, colitis, food allergy

Arrhythmia	Heart Disease- Arteriosclerosis, high cholesterol, high triglycerides
Arthritis- Rheumatoid and Osteoarthritis	Heart Disease- in infants born to magnesium deficient mothers
Asthma	High Blood Pressure
Autism	Hypoglycemia
Auto immune disorders- all types	Impaired athletic performance
Cavities	Infantile Seizure- in children from magnesium deficient mothers
Cerebral Palsy- in children from magnesium deficient mothers	Insomnia
Chronic Fatigue Syndrome	Kidney Stones
Congestive Heart Disease	Lou Gehrig's Disease
Constipation	Migraines- including cluster type
Crooked teeth- narrow jaw- in children from magnesium deficient mothers	Mitral Valve Prolapse
Depression	Multiple Sclerosis
Muscle cramps	PPH- Primary Pulmonary Hypertension
Muscle weakness, fatigue	Raynaud's
Myopia- in children from magnesium deficient mothers	SIDS- Sudden Infant Death Syndrome
Obesity- especially obesity associated with high carbohydrate diets	Stroke
Osteoporosis- just adding magnesium reversed bone loss	Syndrome X- insulin resistance
Parkinson´s Disease	Thyroid disorders- low, high and auto-immune; low magnesium reduces T4
PMS- including menstrual pain and irregularities	

Instead of ordering large batteries of tests, doctors' first reflex action should be to put their patients immediately on magnesium. If they had done that earlier as a preventive action, the chances are much greater they would not have their patients sitting in their offices with so many complaints. It should become clear that as breathing is vital to life, magnesium is fundamental health, as a vital food, and in cases of deficiency, as an essential medicine.

The chances of magnesium not being exceedingly helpful is very low, not only because of the massive deficiencies evident in all segments of the population today, but because it is exceptionally supportive of all aspects of cell physiology in an extremely safe way.

Because the allopathic medical industrial complex is dragging its feet on using magnesium as a universal medicine, the entire alternative medical sector has a chance to make even bigger inroads in the health field. Americans' dissatisfaction with traditional health care is seen in the more than $27 billion they spend annually on alternative and complementary medicine. The numbers continue to grow for reasons that have to do with increasing distrust of mainstream medicine and how it is maintaining a blind eye to medical logic and sound scientific principles. Alternative practitioners can provide a profound benefit to their patients by integrating magnesium into the foundations of their medical and healthcare practices.

Eventually more mainstream doctors, and the schools that train them, will get the idea, or like the dinosaurs they will pass into oblivion — relics of an age of medical ignorance and arrogance. Men and women, whose primary allegiances are to the pharmaceutical companies, have an exceedingly difficult time listening to anything that challenges the basic practice and philosophy of allopathic medicine.

Though this might feel like war talk we must remember that millions of people around the world are dying of iatrogenic (i.e., induced inadvertently via medical diagnosis, surgery, or treatment) death each year and uncounted millions suffer from iatrogenic diseases.

In a medically sane world it should be our hope that this division between allopathic and alternative medicine would not be so great or that they would meld together with mutual respect and knowledge. Yet it is more like a war and people are dying at a rate not equaled in any war in history. Unless this changes, in the future we will have to add to the already astronomically high iatrogenic statistics, a huge group that would not be sick and dying in the first place if it were not for the fact that doctors are neglecting their magnesium needs, when today, there is no need to, and in fact, the benefits are way too obvious to ignore.

On February 5, 2006 the New York Times suggested we declare "war on diabetes" and the first battle should be fought at the Food and Drug Administration (FDA). The FDA is, in effect, the headquarters of all that is wrong and hurtful in medicine. It has successful censored every claim that dietary ingredients treat disease, regardless of the proof in support of the claim.[26] The FDA only allows *drug companies* to make claims of treatment. It also only acknowledges drugs (not food) as the only effective response to diseases. In effect, the FDA protects a monopoly for pharmaceutical companies at the expense of the lives of people.

Magnesium walks a fine line between its perception as a dietary ingredient (let's call it a *food*), and a medicinal agent because it is a powerful medicine already used in emergency rooms. Yet, magnesium would not be needed nearly as much in the emergency room if it was in our foods, and we were putting sufficient amounts in our bodies "the old fashion way."

White blood cells and other immune components are sensitive to malnutrition, especially when it comes to magnesium.

In medicine, arrogance translates into massive death and suffering. It is not a time to be nice, or to mince our words on the state of medical truth these days. The truth is that the need for preventive magnesium medicine has never been greater and the cost of doctors' resistance to coming around to this point of view is unconscionable.

Magnesium affects lung function and indirectly influences respiratory symptoms.

Magnesium is at the center of many processes that are important to body-energy metabolism, immune function, and muscle and nerve function. It is a vital component in the body's physiology, and the essential preventive medicine for the 21st century.

While there are a lot of unknowns when it comes to staying healthy and little, if any, evidence that some therapies that people use today actually work, this is not the case at all when it comes to magnesium. The evidence is ironclad and according to the U.S. Department of Agriculture *1994 Continuing Survey of Food*, intakes by individuals, the mean magnesium intake by males age 9 and older was 323 mg/day — far below today's RDA of 420 mg/day. Similarly, for women older than 9, the mean intake was 228 mg/day — again, significantly below the RDA of 320 mg/day. The proportion of individuals consuming their RDA of magnesium is very low thus it is essential that magnesium be used supplementally to bring levels up to proper balance, and as a primary medical intervention.

References

1   The New York Times February 2, 2006

2   *Dietary Magnesium, Potassium, Sodium, and Children's Lung Function.* Gilliland et al. Am. J. Epidemiol..2002; 155: 125-131.

3   For more information, see: www.mgwater.com/wellness.shtml

4   See: www.foodnavigator.com/news/ng.asp?n=65593-enzyme-nutritional-obese

5   Addressing the assertion that changes in the methods of measuring the composition of food cannot account for the huge fall in nutrient content, Dr Tim Lobstein said, "Minerals are easy to detect and measure and have been since the 19th century. It is almost impossible that methods have changed so much that it would explain the huge difference between these figures. One of the key arguments is that today's agriculture does not allow the soil to enrich itself, but depends on chemical fertilizers that don't replace the wide variety of nutrients (that) plants and humans need."

6   See: www.mgwater.com/wellness.shtml

7   Toward practical prevention of Type II diabetes. McCarty MF. Med Hypotheses 2000;54:786-793.

8   Migraine Awareness Group: a National Understanding for Migraineurs

9   Harrison, Tinsley R. *Principles of Internal Medicine.* 1994, 13th edition, McGraw-Hill, pp. 1106-15 and pp. 2434-35

10   Shechter, Michael, et al. The rationale of magnesium supplementation in acute myocardial infarction: a review of the literature. *Archives of Internal Medicine*, Vol. 152, November 1992, pp. 2189-96

11   Ott, Peter and Fenster, Paul. Should magnesium be part of the routine therapy for acute myocardial infarction? *American Heart Journal*, Vol. 124, No. 4, October 1992, pp. 1113-18

12   Dubey, Anjani and Solomon, Richard. Magnesium, myocardial ischaemia and arrhythmias: the role of magnesium in myocardial infarction. *Drugs*, Vol. 37, 1989, pp. 1-7

13   England, Michael R., et al. Magnesium administration and dysrhythmias after cardiac surgery. *Journal of the American Medical Association*, Vol. 268, No. 17, November 4, 1992, pp. 2395-2402

14   Yusuf, Salim, et al. Intravenous magnesium in acute myocardial infarction. *Circulation*, Vol. 87, No. 6, June 1993, pp. 2043-46

15   Woods, Kent L. and Fletcher, Susan. Long-term outcome after intravenous magnesium sulphate in suspected acute myocardial infarction: the second Leicester Intravenous Magnesium Intervention Trial (LIMIT-2). *The Lancet*, Vol. 343, April 2, 1994, pp. 816-19

16   Eisenberg, Mark J. Magnesium deficiency and sudden death. *American Heart Journal*, Vol. 124, No. 2, August 1992, pp. 544-49

17   Supplemental dietary potassium reduced the need for antihypertensive drug therapy. *Nutrition Reviews*, Vol. 50, No. 5, May 1992, pp. 144-45

18   Ascherio, Alberto, et al. A prospective study of nutritional factors and hypertension among US men. *Circulation*, Vol. 86, No. 5, November 1992, pp. 1475-84

19   Witteman, Jacqueline C.M., et al. Reduction of blood pressure with oral magnesium supplementation in women with mild to moderate hypertension. *American Journal of Clinical Nutrition*, Vol. 60, July 1994, pp. 129-35

20   Geleijnse, J.M., et al. Reduction in blood pressure with a low sodium, high potassium, high magnesium salt in older subjects with mild to moderate hypertension. *British Medical Journal*, Vol. 309, August 13, 1994, pp. 436-40

21   Manz, M., et al. Behandlung von herzrhythmusstorungen mit magnesium. *Deutsche Medi Wochenschrifte*, Vol. 115, No. 10, March 9, 1990, pp. 386-90

22   Iseri, Lloyd T., et al. Magnesium therapy of cardiac arrhythmias in critical-care medicine. *Magnesium*, Vol. 8, 1989, pp. 299-306

23   The study, dubbed the Magpie Trial, was a large international effort aimed at discovering the effects of magnesium sulfate on women with preeclampsia and their children. Close to 10,000 women with preeclampsia from 33 developed and developing countries were involved. Roughly half of the women were randomly assigned to receive magnesium sulfate while the other half received a placebo. Use of magnesium sulfate resulted in a 58% decrease in risk of eclampsia compared to use of the placebo. This translates to 11 fewer women in 1,000 suffering from eclampsia. The preventive effect of magnesium was consistent regardless of the severity of the preeclampsia, the stage of pregnancy, whether an anticonvulsant had been given prior to the trial, and whether the woman had delivered before entry into the trial. Women receiving magnesium sulfate also had a 45% lower risk of death than women receiving the placebo. There appeared to be no difference in the risk of fetal or infant death related to the use of either the drug or the placebo.

24   See: www.mgwater.com/wellness.shtml

25   *Harvard Heart Letter*. November 2002

26   Between 1992 and 1996, FDA prohibited companies that sell folic acid from telling women of childbearing age that .4 mg of folic acid daily before pregnancy could reduce the incidence of neural tube defects (including spina bifida and encephaly) by 40%. FDA's censorship contributed to a preventable 10,000 neural tube defect births.

Between 1994 to 2000, FDA prohibited companies that sell omega-3 fatty acids from telling Americans that those fatty acids found in fish oil could reduce the risk of coronary heart disease by as much as 50%. FDA's censorship contributed to a preventable 1.8 million sudden death heart attacks.

Between 2000 and the present, FDA prohibits companies that sell saw palmetto extract (the fruit of the dwarf American palm tree) from telling Americans that saw palmetto reduces enlarged prostates and relieves related symptoms. Approximately 50% of all men age 50 and older suffer from enlarged prostates and are denied access to this information.

Between 2000 and the present, FDA prohibits companies that sell glucosamine and chondroitin sulfate from telling Americans that those dietary ingredients treat osteoarthritis and relieve osteoarthritic pain and stiffness. Approximately 20 million Americans suffer needlessly from osteoarthritis.

6

||||||||||||||||||||||||||||||||||||||||||||||||||||||||||||||||||||||||||||

Magnesium, Selenium, and Zinc in Cancer Prevention

Since ending one's life with cancer is not pleasant
it behooves all of us to be concerned with its prevention.

It certainly is time to get serious about cancer prevention, with the disease predicted to surge in the next 15 years. The Association for International Cancer Research (AICR) said that if current trends continue, the number of people developing cancer was set to rise at an "alarming" rate. The World Health Organization predicts that cases of cancer will increase by up to 50% worldwide by 2020.

The weight of evidence based on the findings of wildlife biologists, toxicologists, and epidemiologists clearly indicates that the world's populations are being exposed to a host of chemical contaminants that have recently passed over an invisible barrier and is now more dangerous and threatening than any combination of viruses.

Chronic disease is the number one killer in the United States, accounting for about four out of five deaths in America each year. According to the Physicians for Social Responsibility (PSR) about 100 million Americans, more than one-third of the population, suffer from some form of chronic disease like asthma, diabetes, cancer, heart and kidney disease or arthritis.

Cancer is the second leading cause of death, exceeded only by heart disease. *Among children ages 1 to 14, cancer is now the leading cause of death by disease.* At current rates, invasive cancer will be diagnosed in half of all men and in one in three women in their lifetime.

More than 1.3 million new cases of invasive cancer will have been diagnosed in 2006, meaning that approximately 1,500 Americans will die of the disease everyday. "Whether it is cancer or autism that is affecting our families and showing up in our examination rooms, the growing rates of chronic disease compel us to search for clues and answers to determine the true causes of these increasingly prevalent illnesses," says the PSR.

Almost 100% of these people are
suffering from chemical poisoning.

"With rates of cancer incidence rising, mortality rates not falling, and an ever increasing arsenal of high-tech scanners, radiotherapy equipment, and chemotherapeutic drugs being directed in what sometimes appears as a losing battle, there is no more emotive nor scientifically charged issue than cancer," writes Dr.

Sandra Goodman. Along with the rest of the allopathic medical establishment, the last thing oncologists want to admit is that the population is suffering from poisoning from hundreds of carcinogenic compounds and that this, in large part, is a great part of what is driving the escalating epidemic in cancer.

In March of 2004, the U.S. federal government issued an unusually detailed alert to the nation's 5.5 million health care workers: The powerful drugs used in chemotherapy can themselves cause cancer and pose a risk to nurses, pharmacists and others who handle them. Four years in the making, the alert was issued by the National Institute for Occupational Safety and Health (NIOSH).

Chemotherapy — the use of potent drugs to kill cancerous cells — is more than 60 years old. The first such drugs were nitrogen mustards, originally developed as chemical warfare agents. Modern chemotherapy drugs are so strong that they can cause secondary cancers in patients; to a healthy person, they're poison. Most health care workers are clueless about how toxic these agents really are. When oncologists use radiation and chemotherapy, they are using agents that cause cancer to treat the disease.

A Harvard thesis has shown a connection
between water fluoridation and a 700% increase
in osteosarcoma in young men if they are exposed to
fluoridated water during their 6th to 8th years.[1]
Dr. Paul Connett

Prevailing medical response protocols for cancer patients today give only tacit attention to nutrition as a factor in treatment and prevention. This clearly puts the cancer cure advocacy industry and health officials in clear opposition to many distinguished research scientists.

Though data from thousands of studies support taking a less toxic, more holistic approach to treatment of the disease, the majority of allopathic specialists still hold little more than a patronizing view of patients who wish to use a nutrition-based, immune system enhancing strategy in their cancer treatment.

Vociferous and over-zealous protestation by oncologists, who are convinced that nutrition in cancer treatment is inherently worthless and tantamount to quackery, is embarrassing to the institution of medicine. It exposes many of these doctors as little more than medical automatons, who do only as they have been taught, adhering to established medical orthodoxy, extracting money and making profits for drug companies on the backs and *lives* of cancer patients.[2]

Most mainstream physicians are unaware of the extensive depth of evidence about nutrients preventing and alleviating many deadly diseases. Again, it is because they have not been taught to consider, much less to respect, "alternative", meaning non-pharmaceutical based, approaches to the treatment of disease. The emergence of integrative medicine programs, which was pioneered at the University of Arizona under the tutelage of Dr. Andrew Weil, shows that this trend is changing.

However, it is a trend in need of acceleration, as today, needless suffering remains the rule, rather than the exception in the prevention and treatment of cancer. Nutritional sufficiency, and one of its major mineral components, magnesium, must be plugged in to the solution. Until this happens, we'll only be massaging the status quo.

Selenium

Even according to the National Foundation for Cancer Research, the value of minerals as part of an anticancer diet is frequently overlooked. However, minerals can play a vital role in fighting cancer. In addition to magnesium, a prime example is the mineral selenium, which has powerful antioxidant properties.

Selenium (Se) is an essential micro-nutrient with important biological and biochemical functions in organisms because of its unique antioxidant properties and its ability to regulate thyroid gland metabolism. It is well known that selenium is an antagonist that moderates the toxic effects of many heavy metals such as arsenic, cadmium, mercury, and lead in organisms.

Data suggests that a diet rich in selenium protects against cancers of the stomach, breast, esophagus, lung, prostate, colon, and rectum. According to Dr. Harold Foster, death rates in the USA for cancer are lower when blood selenium levels are high. One important study found that high blood levels of selenium are associated with a four- to fivefold decrease in the risk of prostate cancer.

Scientists at Stanford University studied 52 men who had prostate cancer and compared them to 96 men who did not.[3] One surprising finding was that blood levels of selenium generally decreased with age. It is well known that the risk of prostate cancer increases dramatically as a man advances in age.

Those who have studied geographical differences have seen that in low-selenium regions, higher death rates occurred from malignant lymphomas and cancers of the tongue, esophagus, stomach, colon, rectum, liver, pancreas, larynx, lung, kidneys and bladder.

In addition, cancer patients with low selenium levels tend to have a wider spread of the disease, more recurrences and die sooner.[4]

In China, where the selenium levels in the soils varies much more dramatically than in the United States and the population is less mobile, an ecological study in 1985 showed dramatic results in linking cancer with selenium deficiencies.

Dr. Shu-Yu Yu measured the selenium content of blood stored in blood banks in thirty different regions in China, and classified the regions as high selenium, medium selenium, and low selenium. They then compared death rates from cancer to the selenium rates and found there was an exact correlation. In the low selenium classification, three times as many people died from cancer as in the high selenium classification.

The West African country of Senegal is dominated by high concentrations of selenium in the soil and thus in their foods. As expected, we find that Senegalese males had the world's lowest rates for cancer of the trachea, bronchus and lung; stomach and colon; the fourth lowest for prostate cancer and sixth lowest for esophageal cancer.

Senegalese women had the lowest incidence of cancers of the trachea, bronchus, lung, esophagus, stomach and colon and second lowest for breast cancer and fifth lowest for cancer of the uterus.

There is no doubt that selenium is essential for human health and that these elements will most likely protect against cancer and other diseases. It is also no doubt that the levels of reduction of these essential minerals, in the content of the foods that the vast majority of the population consumes, have reached systemic levels.

For these reasons, people in regions which are naturally rich in selenium tend to live longer. Selenium, especially when used in conjunction with vitamin C, vitamin E and beta-carotene, works to block chemical reactions that create free radicals in the body (which can damage DNA and cause degenerative change in cells, leading to cancer). Selenium also binds strongly with mercury, protecting us from its damaging effects.

Selenium helps stop damaged DNA molecules from reproducing, meaning that it acts to prevent tumors from developing. "It (selenium) contributes towards the death of cancerous and pre-cancer cells. Their death appears to occur before they replicate, thus helping stop cancer before it gets started," says Dr. James Howenstine in *A Physician's Guide to Natural Health Products That Work.*

A 1996 study by Dr. Larry Clark of the University of Arizona showed just how effective selenium can be in protecting against cancer. In a study of 1,300 older people, the occurrence of cancer among those who took 200 micrograms of selenium daily for about seven years was reduced by 42% compared to those given a placebo.

Cancer deaths for those taking selenium were cut almost in half, according to the study that was published in the *Journal of the American Medical Association* on December 25, 1996. In addition, the people who had taken selenium had 63% fewer prostate cancers, 58% fewer colorectal cancers, 46% fewer lung cancers and overall 37% fewer cancers. *Selenium was found to reduce the risk of lung cancer to a greater degree than stopping smoking.*[5]

Magnesium

It is generally accepted that a higher magnesium intake in the drinking water is associated with reduced cancer incidence and reduced frequency of cardiac infarction.

Information is scarce about the relationship between cancer and magnesium but researchers from the School of Public Health at the University of Minnesota have just concluded that diets rich in magnesium reduced the occurrence of colon cancer.[6] A previous study from Sweden[7] reported that women with the highest magnesium intake had a 40% *lower* risk of developing the cancer than those with the lowest intake of the mineral.

Preliminary data also suggests a relationship between low intake of magnesium and kidney cancer.

Several studies have shown an increased cancer rate in regions with low magnesium levels in soil and drinking water as well. In Egypt the cancer rate was only about 10% of that in Europe and America. In the rural fellah it was practically non-existent. The main difference was an extremely high magnesium intake of 2.5 to 3 g in these cancer-free populations, ten times more than in most western countries.[8]

Dr. Seeger and Dr. Budwig in Germany have shown that cancer is mainly the result of a faulty energy metabolism in the powerhouses of the cells, the mitochondria. ATP and most of the enzymes involved in the production of energy *require* magnesium. A healthy cell has high magnesium and low calcium levels. The problem that comes with low magnesium (Mg) levels is the calcium builds up inside the cells while energy production decreases as the mitochondria gradually calcify.

"Magnesium is critical for all of the energetics of the cells because it is absolutely required that the magnesium be bound (chelated) by ATP (adenosine triphosphate), the central high energy compound of the body. ATP without bound magnesium cannot create the energy normally used by specific enzymes of the body to make protein, DNA, RNA, transport sodium or potassium or calcium in and out of cells, nor to phosphorylate proteins in response to hormone signals, etc. In fact, ATP without enough magnesium is non-functional and leads to cell death.

"Bound magnesium holds the triphosphate in the correct stereochemical position so that it can interact with ATP using enzymes. Magnesium also polarizes the phosphate backbone so that the 'backside of the phosphorous' is more positive and susceptible to attack by nucleophilic agents such as hydroxide ions or other negatively charged compounds.

"Bottom line, magnesium at critical concentrations is essential to life," says Dr. Boyd Haley, who also asserts strongly that, "All detoxification mechanisms have as the bases of the energy required to remove a toxicant the need for magnesium and ATP to drive the process. There is nothing done in the body that does not use energy and without magnesium this energy can neither be made nor used." Detoxification of carcinogenic chemical poisons is essential for people want to avoid the ravages of cancer. The importance of magnesium in cancer prevention should *not continue being* underestimated.

The School of Public Health at the Kaohsiung Medical
College in, Taiwan, found that magnesium also exerts a
protective effect against gastric cancer, but only
for the group with the highest levels.[9]

Among other effects, magnesium improves the internal production of defensive substances, such as antibodies and considerably improves the operational activity of white, granulozytic blood cells (shown by Delbet with magnesium chloride), and contributes to many other functions that insure the integrity of cellular metabolism.

In 1961, Dr. Hans A. Nieper, introduced cardiac therapy based on magnesium aspartate. He was surprised to observe that hardly any new cancer occurrences appear in the group of patients so treated. The rate of new cancerous diseases with long-term magnesium therapy was reported to be less than 20% of the frequency otherwise expected.

In an uncontrolled trial, researchers in the UK found that intravenous magnesium relieves neuropathy pain in patients with cancer.[10] Magnesium acts as a noncompetitive antagonist of the N-methyl-D-aspartate receptor, which has been implicated in the transmission of pain, according to Dr. Vincent Crosby and colleagues at Nottingham City Hospital.

It is known that carcinogenesis induces magnesium distribution disturbances, which cause magnesium mobilization through blood cells and magnesium depletion in non-neoplastic tissues. Magnesium deficiency seems to be carcinogenic, and in cases of solid tumors, a high level of magnesium inhibits carcinogenesis.[11] Both carcinogenesis and magnesium deficiency increase the plasma membrane permeability and fluidity.

Zinc

No modern chemotherapy includes zinc adjuvant even though zinc serum levels are usually low in leukemic children.

Epidemiologic studies suggest that zinc deficiency may be associated with increased risk of cancer.[12] Zinc supplementation is associated with decreased oxidative stress and improved immune function, which may be among the possible mechanisms for its cancer preventive activity.

Zinc is essential for health. It is needed for the enzymes that regulate cell division, growth, wound healing, and proper functioning of the immune system. Zinc is an essential co-factor in a variety of cellular processes including DNA synthesis, behavioral responses, reproduction, bone formation, growth and wound healing.

Zinc is a component of insulin and it plays a major role in the efficiency of most of the functions of the body. Zinc is necessary for the free-radical quenching activity of superoxide dismutase (SOD), a powerful antioxidant enzyme which breaks down the free-radical superoxide to form hydrogen peroxide.

Zinc is required for the proper function of T-lymphocytes. The mineral also plays a role in acuity of taste and smell. And zinc is required for proper functioning of genetics, immunity, formation of red blood cells, organ, muscle and bone function, cell membrane stability, cell growth, division, differentiation and genetics. Importantly, zinc is vital for the metabolism of vitamin A.

A paper by Dr. Mei and colleagues at the Anhui Medical University, Hefei, China, suggests that some aspects of immune function can be enhanced by treatment with zinc. The authors state that it would be "reasonable to expect that zinc is instrumental in restoring failing immuno-competence of cancer patients. Mei studied the influence of zinc and selenium-zinc upon the immune function (T-cells, granulocytes and NK cells) of cancer patients. The results showed that immune response was strengthened.[13]

Leukemic cells contain much less zinc than normal lymphocytes, suggesting an error in zinc metabolism, which appears correctable with zinc treatment. Zinc also is known to have some beneficial interactions with chemotherapy drugs. In one recent case, upon noting low blood levels of zinc in a 3-year-old 11.3 kg girl, zinc was administered at the rate of 3.18 mg/kg body weight per day from the start of chemotherapy through the full 3 years of maintenance therapy. Dosage was split with 18 mg given at breakfast and 18 mg zinc with supper. The result was a bone marrow remission from 95+% blast cells to an observed zero blast cell count in both hips within the first 14 days of treatment which never relapsed.[14]

Dr. Mathais Rath wrote that "500 years ago, the Roman church was making billions of Thaler (early dollars) by selling indulgences, an imaginary "key to heaven" for its believers. Then the fraud scheme collapsed and with it much of the power of the church. Today, the pharma business uses the same fraud scheme. It tries to sell the 'key to health' to millions of people and takes away billions of dollars in return for an illusion: the deception that the pharmaceutical industry is interested in your health."[15]

The cancer industry is among the most aggressive areas in the medical arena and the FDA will swoop in with automatic weapons into doctors' offices if they step out of line with acknowledged oncology protocols. Medical fascism is perhaps at its worst in the cancer area with oncologists insisting that poisoning patients with chemicals and radiation is the *only and best* way to go.

Minerals, not drugs, and certainly not radiation or chemicals, are essential for life and health and provide the keys for the prevention and cure of cancer. Minerals (in the form of cesium chloride)[16] also provide a reasonably safe way to treat advanced stage four cancer without resorting to the slash and burn tactics of radiation and chemotherapy.

It is a frightening world when it comes to cancer and that fright is made much worse by the medical authorities who insist their way is the only way. Not only do they let the ball drop by not informing us about how to avoid cancer, but if we get it, they will use therapies that are as dangerous to live healthy cells as they are to the cancer cells. Cesium chloride on the other hand aggressively kills cancer cells by turning their internal chemistry alkaline.

References

1   Dr. Paul Connett posted on 20 July 2005 at 4:55 am. I am really surprised that *Medical News Today* published the puff piece from the American Dental Association about their celebration of 60 years of fluoridation, but missed the real news from last week. This was the revelation carried by the Washington Post and the Associated Press (July 13, 2005) that a Harvard thesis has shown a connection between water fluoridation and a 700% increase in osteosarcoma in young men if they are exposed to fluoridated water during their 6th to 8th years. Particularly disturbing is the information that the thesis adviser, Porfessor Cheser Douglass, who is also a consultant to Colgate, has covered up these results in talks to the public and in a report to his funding agency. Both the NIEHS and Harvard University are investigating his conduct.

2   Some doctors just take it too far. Dr. Michael A. Rosin, for instance, was accused of falsely diagnosing patients with skin cancer and operating on them unnecessarily. He was recently ordered by federal authorities to refer all patients with confirmed or suspected skin cancer to other doctors instead of treating them himself. The order says Rosin, 54, poses "an immediate and serious danger to the health, safety and welfare of the public." He was found guilty by jury trial.

3   *The Journal of Urology* {2001;166:2034-8}. December issue.

4   Foster HD. "Landscapes of Longevity: The Calcium-Selenium-Mercury Connection in Cancer and Heart Disease," *Medical Hypothesis*, Vol. 48, pp 361-366, 1997.

5   Clark LC. The epidemiology of selenium and cancer. *Fed Proc* 1985; 44:2584-2590.

6   *American Journal of Epidemiology* (Vol. 163, pp. 232-235)

7   *Journal of the American Medical Association*, Vol. 293, pp. 86-89

8   MAY 19, 1931, Dr. P. Schrumpf-Pierron presented a paper entitled "On the Cause Of the Rarity of Cancer in Egypt," which was printed in the *Bulletin of the Academy of Medicine*, and the *Bulletin of the French Association for the Study of Cancer* in July, 1931. www.mgwater.com/rod02.shtml

9   Yang CY et al. Jpn *J Cancer Res*.1998 Feb;89 (2):124-30. Calcium, magnesium, and nitrate in drinking water and gastric cancer mortality.

10   Reuters Health, Feb. 10, 2000 AND the *Journal of Pain and Symptom Management,* Jan. 2000; 19:35-39

11   Durlach J, Bara M, Guiet-Bara A, Collery P. Relationship between magnesium, cancer and carcinogenic or anticancer metals. *Anticancer Res.* 1986 Nov-Dec;6(6):1353-61.

12   Prasad AS, Kucuk O. Zinc in cancer prevention. Department of Medicine (Division of Hematology-Oncology), Barbara Ann Karmanos Cancer Institute, Wayne State University School of Medicine Detroit, MI 48201, USA. *Cancer Metastasis Rev.* 2002;21(3-4):291-5.

13   Mei W et al. Study of immune function of cancer patients influenced by supplemental zinc or selenium-zinc combination. *Biol Trace Elem Res*; 28(1):11-9. Jan 1991.

14   George Eby. Treatment of acute lymphocytic leukemia using zinc adjuvant with chemotherapy and radiation – a case history and hypothesis. *Medical Hypotheses* (2005) 64, 1124–1126

15   See: www4.dr-rath-foundation.org/PHARMACEUTICAL_BUSINESS/ pharmaceutical_industry.htm

16   USE OF CESIUM CHLORIDE TO CURE MALIGNANCIES, www. newswithviews.com/Howenstine/james14.htm

7

||||||||||||||||||||||||||||||||||||||||||||||||||||||||||||||||

Magnesium – Antioxidant Status – Glutathione

The involvement of free radicals in tissue injury is induced by magnesium deficiency. Magnesium deficiency has been associated with the production of reactive oxygen species, cytokines, and eicosanoids[1], as well as vascular compromise *in vivo*. Although magnesium deficiency induced inflammatory change occurs in "chronic" cases *in vivo*, acute deficiency may also affect the vasculature (blood vessels) and consequently, predispose endothelial[2] cells (EC) to perturbations associated with chronic magnesium deficiency.

As oxyradical production is a significant component of chronic magnesium deficiency, we examined the effect of acute states on endothelial cell oxidant production *in vitro*. In addition, we determined EC; pH, mitochondrial function, lysosomal integrity and general cellular antioxidant capacity. Decreasing Mg_2+

($<$ or $=$ 250microM) significantly increased EC oxidant production relative to control Mg_2+ (1000microM). Magnesium deficiency induced oxidant production, occurring within 30 minutes, was attenuated by EC treatment with oxyradical scavengers and inhibitors of eicosanoid biosynthesis.[3] Coincident with increased oxidant production were reductions in intracellular glutathione (GSH) and corresponding EC alkalinization. These data suggest that acute magnesium deficiency is sufficient for induction of EC oxidant production, the extent of which may determine, at least in part, the extent of EC dysfunction/injury associated with chronic magnesium deficiency.

Magnesium deficiency causes an accumulation of oxidative products in the heart, liver, kidney, skeletal muscle tissues and in red blood cells.[4] Magnesium is a crucial factor in the natural self-cleansing and detoxification responses of the body. It stimulates the sodium potassium pump on the cell wall and this initiates the cleansing process in part because the sodium-potassium-ATPase pump regulates intracellular and extracellular potassium levels.

Cell membranes contain a sodium/potassium ATPase, a protein that uses the energy of ATP to pump sodium ions out of the cell, and potassium ions into the cell. The pump works all of the time, like a bilge pump in a leaky boat, pumping potassium and sodium in and out, respectively.

"ATP production is essential for every cell to have an ample supply to deal with the challenges of metal overload, as it is required to even permit the cell to keep pumping out calcium. Lack of ATP then is the underlying cause of abnormal calcification of tissues," writes Dr. Garry Gordon, of Gordon Research Institute, a leading authority in oral chelation therapies.

Potassium regulation is of course crucial because potassium acts as a counter flow for sodium's role in nerve transmission. The body must put a high priority on regulating the potassium of the blood serum and this becomes difficult when magnesium levels become deficient.[5]

Because of these crucial relationships, when magnesium levels become dramatically deficient we see symptoms such as convulsions, gross muscular tremor, atheloid movements, muscular weakness, vertigo, auditory hyperacusis, aggressiveness, excessive irritability, hallucinations, confusion, and semi-coma. Magnesium deficiency causes the body to lose potassium. Magnesium and potassium inside the cell assist oxidation, and sodium and calcium outside the cell wall help transmit the energy produced. The healthy cell wall favors intake of nutrients and elimination of waste products.

Magnesium protects cells from aluminum, mercury, lead, cadmium, beryllium and nickel, which explains why re-mineralization is so essential for heavy metal detoxification and chelation.

Magnesium protects the cell against oxyradical damage and assists in the absorption and metabolism of B vitamins, vitamin C and E, which are anti-oxidants important in cell protection. Recent evidence suggests that vitamin E enhances glutathione levels and may play a protective role in magnesium deficiency-induced cardiac lesions.[6]

In general, magnesium is essential for the survival of our cells but takes on further importance in this current age of toxicity where our bodies are being bombarded on a daily basis with heavy metals. Magnesium is especially needed to protect the brain from the toxic effects of chemicals. It is highly likely that low total body magnesium contributes to heavy metal toxicity in children and is a strong participant in the etiology of learning disorders and autism.

Without sufficient magnesium, the body accumulates toxins and acid residues, and thusly degenerates more rapidly and ages prematurely. Recent research has pointed to low glutathione levels being responsible for children's vulnerability to mercury poisoning from vaccines.[7] It seems reasonable to assume that low levels of magnesium would also render a child vulnerable. And in fact glutathione requires magnesium for its synthesis.[8] Glutathione synthetase requires γ-glutamyl cysteine, glycine, ATP, and magnesium ions to form glutathione.[9]

In magnesium deficiency, the enzyme γ-glutamyl transpeptidase is lowered.[10] Data demonstrates a direct action of glutathione both *in vivo* and *in vitro* to enhance intracellular magnesium and a clinical linkage between cellular magnesium, GSH/GSSG ratios, and tissue glucose metabolism.[11]

Magnesium deficiency causes glutathione loss, which is not affordable because glutathione helps to defend the body against damage from cigarette smoking, exposure to radiation, cancer chemotherapy, and toxins such as alcohol and just about everything else.

According to Dr. Russell Blaylock, low magnesium is associated with dramatic increases in free radical generation as well as glutathione depletion and this is vital since glutathione is one of the few antioxidant molecules known to neutralize mercury.[12] Thus, sadly, children receiving thimerosal containing vaccines are sitting ducks to mercury when both magnesium and glutathione levels are low.

Also under the shadow of magnesium deficiency, too much Nitric Oxide (NO) is produced which in turn may react with superoxide to form a very damaging compound peroxynitrite. Low magnesium levels can induce such excessive NO production that even the glutathione in the red blood cells is damaged. This provides one explanation for why magnesium protects the arteries.[13]

"For every molecule of pesticide that your body detoxifies, you throw away or use up forever, a molecule of glutathione, magnesium and more," says Dr. Sherry Rogers who goes on to say that, "Your body uses nutrients to make this glutathione and it uses up energy as well. Every time we detoxify a chemical, we use up, lose, throw away forever, a certain amount of nutrients."

Magnesium permits calcium to enter a nerve cell to allow electrical transmission along the nerves to and from the brain. Even our thoughts, via brain neurons, are dependent on magnesium.

Dr. Carolyn Dean

When dealing with autism spectrum and other neurological disorders in children it is important to know the signs of deficient magnesium: restlessness, can't keep still, body rocking, grinding teeth, hiccups, noise sensitive, poor attention span, poor concentration, irritable, aggressive, ready to explode, easily stressed. When it comes to autistic children today we need to assume significant magnesium deficiency for multiple reasons.

•   The foods they are eating are stripped of magnesium because of soil deficiencies in vital minerals.
•   The diets of many children consists largely of highly processed junk foods that do not provide real nutrition to the body.

• Children on the autistic spectrum are not absorbing the minerals they need even when present in the gut. Magnesium absorption is dependent on intestinal health, which is compromised in leaky gut syndrome.

• The oral supplements doctors rely on are not easily absorbed.

Magnesium and copper are important modulators of NMDA-receptor activity. Recent data indicate that disturbances of glutamatergic transmission (especially via NMDA-receptor) are involved in the pathogenesis of mood disorders.

Magnesium deficiency is related to a variety of psychological symptoms, especially depression. There are many reports indicating significant changes in blood levels of magnesium or copper during a depressive episode. Moreover magnesium exhibits antidepressant-like and anxiolytic-like effects in animal models of depression, in rodents.[14]

Evidence is mounting that deficient levels of magnesium contribute to the heavy metal deposition in the brain that precedes Parkinson's disease, Multiple Sclerosis and Alzheimer's disease.

Many of the symptoms of Parkinson's disease can be overcome with high magnesium supplementation. In a trial with thirty epileptics, 450 mg of magnesium supplied daily successfully controlled seizures. Another study found that the lower the magnesium blood levels the more severe was the epilepsy.

Magnesium works best in combination
with vitamin B$_6$ and zinc.

Because of its nerve and muscle support, magnesium is helpful for nervousness, anxiety, insomnia, depression, and muscle cramps. Thus magnesium is also given as part of a treatment for autism or hyperactivity in kids.

Dr. Bernard Rimland, of the Autism Research Institute, did extensive research on vitamin B_6 and magnesium many years ago and found, through double-blind placebo-controlled crossover experiments with sixteen autistic children, statistically significant results. Children and adults tend to sleep better when taking magnesium before bed.

References

1   In biochemistry, eicosanoids are a class of oxygenated hydrophobic molecules that largely function as autocrine and paracrine mediators. Eicosanoids derive from 20-carbon polyunsaturated essential fatty acids, most commonly arachidonic acid (AA) in humans.

2   A thin layer of flat epithelial cells that line serous cavities, lymph vessels, and blood vessels.

3   Wiles ME, Wagner TL, Weglicki WB.The George Washington University Medical Center, Division of Experimental Medicine, Washington, D.C., USA. mwiles@nexstar. com Life Sci. 1997;60(3):221-36.

4   Martin, Hélène. Richert, Lysiane. Berthelot, Alain Magnesium Deficiency Induces Apoptosis in Primary Cultures of Rat Hepatocytes.* Laboratoire de Physiologie, et Laboratoire de Biologie Cellulaire, UFR des Sciences Médicales et Pharmaceutiques, Besançon, France. 2003 *The American Society for Nutritional Sciences J. Nutr.* 133:2505-2511, August 2003

5   A magnesium deficiency can cause the body to lose potassium [Peterson 1963][M acIntyre][Manitius], possibly because of a poorly understood effect of magnesium on the efficiency of energy supply to the sodium pump [Fischer].

6   Barbagallo, Mario et al. Effects of vitamin E and Glutathione on Glucose Metabolism: Role of Magnesium; (*Hypertension.* 1999;34:1002-1006.)

7   Enviroonmental Working Group. (www.ewg.org/reports/autism/part1.php)

8   Linus Pauling Institute (http://lpi.oregonstate.edu/infocenter/minerals/magnesium/index.html#function)

9   Virginia Minnich, M. B. Smith, M. J. Brauner, and Philip W. Majerus. Glutathione biosynthesis in human erythrocytes. Department of Internal Medicine, Washington University School of Medicine, *J Clin Invest.* 1971 March; 50(3): 507–513. Abstract: The two enzymes required for de novo glutathione synthesis, glutamyl cysteine synthetase and glutathione synthetase, have been demonstrated in hemolysates of human erythrocytes. Glutamyl cysteine synthetase requires glutamic acid, cysteine, adenosine triphosphate (ATP), and magnesium ions to form γ-glutamyl cysteine. The activity of this enzyme in hemolysates from 25 normal subjects was 0.43±0.04 μmole

glutamyl cysteine formed per g hemoglobin per min. Glutathione synthetase requires γ-glutamyl cysteine, glycine, ATP, and magnesium ions to form glutathione. The activity of this enzyme in hemolysates from 25 normal subjects was 0.19 ± 0.03 μmole glutathione formed per g hemoglobin per min. Glutathione synthetase also catalyzes an exchange reaction between glycine and glutathione, but this reaction is not significant under the conditions used for assay of hemolysates. The capacity for erythrocytes to synthesize glutathione exceeds the rate of glutathione turnover by 150-fold, indicating that there is considerable reserve capacity for glutathione synthesis. A patient with erythrocyte glutathione synthetase deficiency has been described. The inability of patients' extracts to synthesize glutathione is corrected by the addition of pure glutathione synthetase, indicating that there is no inhibitor in the patients' erythrocytes.

10   Braverman, E.R. (with Pfeiffer, C.C.)(1987). The healing nutrients within: Facts, findings and new research on amino acids. New Canaan: Keats Publishing.

11   Barbagallo, M. et al. Effects of glutathione on red blood cell intracellular magnesium: relation to glucose metabolism. *Hypertension*. 1999 Jul;34(1):76-82. Institute of Internal Medicine and Geriatrics, University of Palermo, Italy.

12   See: (www.dorway.org/blayautism.txt)

13   Mak IT; Komarov AM; Wagner TL; Stafford RE; Dickens BF; Weglicki WB Address Department of Medicine, George Washington University Medical Center, Washington, District of Columbia 20037, USA. Source *Am J Physiol*, 1996 Jul, 271:1 Pt 1, C385-90

14   Siwek M., et al. The role of copper and magnesium in the pathogenesis and treatment of affective disorders. *Psychiatr Pol*. 2005 Sep-Oct;39 (5):911-20. Klinika Psychiatrii Doroslych CM UJ

8

Detoxification and Chelation

How many doctors relate the increased accumulation of mercury in the body to deficiencies in magnesium? The answer is, too few. Perhaps in the 1950's and before, there wasn't a reason to make such a connection. Today we can't afford not to.

Today, the cause and cure of many (we're not suggesting all) physical illnesses can be as simple as correcting a magnesium deficiency, in addition to taking active steps to remove accumulations of toxic heavy metals from the body through a process called chelation (pronounced *key-lay-shun*).

Mercury is one such metal that specific steps must be taken to safely remove from the system. Many of the challenges associated with mercury chelation can be reduced with sufficient magnesium intake. Among alternative health practitioners it is widely known that chelation[1] wastes minerals. Yet few have sufficiently given attention to the key mineral whose loss cannot be tolerated without undue, and sometimes grave risk.

Magnesium is very important for phase one detoxification. It, along with other minerals like zinc, displaces toxic heavy metals such as mercury. Magnesium is a crucial factor in the natural self-cleansing and detoxification responses of the body. Thus it is reasonable to assume that sub-optimum levels of magnesium would render a child vulnerable to mercury mobilization during chelation.

Sometimes simply raising your magnesium levels is not enough to restore long-lasting health. Though necessary for detoxing and chelation, magnesium is not a chelator itself. Chelation is a necessary component of any healing or health program because all of us are contaminated with many heavy metals and now even uranium oxide. Though chelation is a big subject, the following will give you a general overview so that you can compare several chelation protocols to help you decide what is best for you with the main options being either the natural route or synthetic drugs that themselves contribute to the toxicity of the body.

The Golden Triangle of Natural Chelation

After discovering that the pristine environments of the world have vanished we now have discovered too late that the same thing has happened to our blood streams and thus cell environments in our bodies.[2] In allowing our corporations to trash the world we have allowed them to trash our bodies[3] with all kinds of chemical pollutants[4] including a long list of dangerous heavy metals.[5] Modern medicine and dentistry have themselves been involved in the gold rush to profit from the use of dangerous chemicals and have contributed greatly to the intense poisoning of our bodies. At the time of our greatest need, when public health has deteriorated especially among the young[6] and elderly, our doctors, dentists and public health care officials have gone blind, deaf and dumb to the problem and refuse to help us in our need. Instead they continue to cling to their covenant with poisons and increase the flow of pollution to our cells.

> *We have almost 20% of our children now in this country chronically ill or disabled. That's a very different situation from what it was 20 or 30 years ago, and there's no explanation given by the public health authorities as to why that is true.*
>
> Barbara Loe Fisher
> National Vaccine Information Center

The World Wildlife Fund (WWF) has been out front in concerning itself with matters that health and medical officials have neglected. In 2004 the organization sponsored a series of blood tests in the UK and the results were astounding. *Every* person tested was contaminated by a cocktail of highly toxic chemicals which were banned from use in the UK during the 1970s and yet continue to pose unknown health risks.[7]

People with chronic illnesses are living testimony to this contamination. In most cases people's limits are reached and exceeded not by an excessive amount of one toxin, allergen or pollutant but by a large oversupply of countless micro-doses, each possibly below what health officials consider dangerous. But together it all explains the great collapse of health and well being in the general public.

The WWF found 70 (90%) of the 78 chemicals they were looking for in the survey. The highest number of chemicals found in any one person was 49, nearly two thirds (63%) of the chemicals looked for. This study provides shocking and damning evidence that people across the world are contaminated with a cocktail of highly toxic chemicals.

Upon studying his personal results, Dr. John Barry of the Green Party said: "As a vegetarian, who eats mainly organic produce, I did not expect my results to indicate a high level of contamination, yet eighteen chemicals were detected in my body. Not a high level in comparison to other results but they are chemicals I did not ask for and certainly do not want." And Michelle Gildernew, Sinn Fein MP said, "Despite following a fairly healthy lifestyle, I was shocked to discover that my blood test revealed a number of chemicals which could be dangerous to health. As a mother, I find it even more disturbing that I could have passed some of these chemicals on to my child during pregnancy."

Fortunately for our children and for us there are natural and semi-natural forms of treatment that can reverse much of the damage of mercury, lead, arsenic and other chemical poisoning. Even uranium can be eliminated with the proper protocol. It's an approach to medicine that is radically different, for instead of adding to the already heavy chemical burden of the body (which almost all allopathic medicines do, including synthetic chelators) it diminishes chemical accumulations, cleans and detoxifies our bodies.

Over the past one hundred years, our species has been engaged in a vast and complicated chemistry experiment. Unbeknownst to anyone, it has been at the expense of our health. Each and every one of us has been guinea pigs in this experiment. However, the experiment failed. We tried to treat diseases by killing the culprits that we thought were causing them. In fact, we treated any "enemy" as something to be killed, as if killing the perpetrator would solve the problem. As a result, we are the most toxic society in history. Our foods have trace toxins from pesticides. Toxins are added to drinking water (to kill bacteria, etc.), which we eventually drink and absorb into our bodies. By trying to make us "safer" from disease by killing, our natural immune systems have been systematically and profoundly compromised, making unsuspecting *us* more susceptible.

Medical protocols that have doggedly looked only to synthetic solutions to the growing problem of chronic disease are fighting a losing battle. This is part of the problem: we're still fighting. We're still killing. We are not strengthening. We're still fearing; everyone and everything. We're still looking for the knockout punch; the silver bullet. But we don't trust Nature's Ways. And in our distrust and fear, we're paying a very high price in life cut short early, liberty abridged, and the pursuit of happiness *not* realized.

Today we face a huge choice for ourselves and our children. We can continue to trust and have faith in our medical and public health officials and mainstream allopathic doctors, who want to continue the experiment, and continue poisoning us and our children. Or we can depart from this medical insanity and start the long work of nourishing ourselves through effective mineral replenishment modalities, and purifying ourselves through detoxification and natural chelation protocols.

Detoxification and chelation are essential medical approaches in the age of toxicity. It must be acknowledged though that chelation approaches that use toxic drugs — as opposed to natural chelation and detoxification agents — could be dangerous and doctors who are using chelation on their young patients would be well served to re-evaluate the use of synthetic chelators.

Metal chelation is a complex and serious matter. You can end up in worse health after chelation than when you started if you are not well-informed and don't proceed carefully under the care of a competent health care practitioner.

I hear many stories involving chelators that are intended for heavy metal removal (such as mercury, cadmium, iron, and lead), as much about the devastating side effects as about the miraculous recoveries. The difference is due, in part, to the methodology of the attending physician, and not necessarily the choice of chelator.

Synthetic chelators can be used to good effect by competent doctors but their dangers are ever present. Natural approaches, on the other hand, are considerably safer. In fact, much can be done independently, that is, unless one is in a disease category that is life threatening.

Some of the more aggressive and toxic chelation procedures are appropriate especially in the cases of acute toxic exposure, stage four cancers, and imminent threat of heart attack or stroke. But even then, when the goal is to remove heavy metals, natural ways can be as quick and as effective as costly and risky intravenous procedures.

Getting Serious About Toxin Removal

In a toxic world an intelligent pharmacology would include affordable, safe substances that facilitate the excretion of toxic metals from the body. Dr. Garry Gordon, a leader in the field of heavy metal detoxification and chelation says, "No one on planet earth is operating at optimal levels without doing something about the toxic metals. Thus the conclusion I draw is that chelation appears a lifetime necessity for all."

Gordon is quick to remind us of important toxic problems like lead. He says, "There is no chelation that can dent the lead levels of bones unless continued for at least 7 years (bone turnover time)."[8] So, if you are betting your patients' health on effective protection you need to be willing to get into chelation and detoxification for the long haul. This is also one of the principle reasons we need natural nontoxic substances.

Using synthetic drugs with their own toxic side effects for long periods of time is not a good idea at all.[9] Even EDTA, which is much less toxic then DMPS and DMSA may not be appropriate for treating low-level lead exposures because it can be toxic in that it increases excretion of some essential metals. EDTA produces substantial diuresis of zinc and a temporary 30-40% decrease in plasma zinc.[10]

In the 21st century the center of pharmacology
needs to be shifted away from medicines that
add to people's already heavy toxic
burdens, to medicines and protocols that
reduce these burdens.

What we need is a unique combination of natural substances, scientifically formulated and tested to form stable complexes with and remove a wide range of toxic heavy metals so they can be readily excreted via multiple pathways in the body. These substances must stimulate and enhance our body's natural endogenous mechanisms for coping with toxic metals and chemicals. This includes stimulating the production of metallothionein (metal binding proteins essential for metal transport), elevating glutathione levels and adding protective essential minerals to the body.

The three most effective, safe and natural substances, when combined, create what I call the golden triangle of natural chelation. They are alpha-lipoic acid (ALA), cilantro, and chlorella (which is not a chelator when used by itself). When used together and supported with strong mineral therapy, they provide the ultimate in safe chelation for a broad array of heavy metals. It was the genius of Dr. Allan Greenburg who brought these three agents together for the first time and tested extensively to prove their effectiveness and safety.

> **Alpha-lipoic acid**: According to Jones and Cherian,[11] an ideal heavy metal chelator should be able to enter the cell easily, chelate the heavy metal from its complex with metallothionein or other proteins, and increase the excretion of the metal without its redistribution to other organs or tissues. According to Dr. Lyn Patrick "ALA satisfies at least two of the above criteria; i.e., absorption into the intracellular environment and complexing metals previously bound to other sulfhydryl proteins. ALA when found unbound in the circulation, is able to trap circulating heavy metals, thus preventing cellular damage caused by metal toxicity. The fact that free ALA crosses the blood brain barrier is significant because the brain readily accumulates lead and mercury, where these metals are stored intracellularly in glial tissue."

> ALA scavenges hydroxyl radicals, singlet oxygen and hypochlorous acid, can remove heavy metals by chelation and regenerates other antioxidants like glutathione, vitamin C, ubiquinol (coenzyme Q10) and indirectly, vitamin E - as such it is

an ideal chelator. A very recent study of children living in the area affected by the Chernobyl disaster also showed that ALA prevents radiation damage.[12] Alpha-lipoic acid and its cousin DHLA have justly been referred to as the "universal antioxidants". They are active in both cell fluids and membranes, they have no serious side effects, are non-carcinogenic and do not interfere with fetal development.

ALA is the oxidized form of dihydrolipoic acid (DHLA). LA contains two thiol (sulfur) groups, which may be oxidized or reduced.

ALA (Oxidized)

Dihydrolipoic acid (DHLA). (Reduced)

Patrick goes on to say, "ALA has been shown to increase both intra-and extracellular levels of glutathione in cell cultures, human erythrocytes, glial cells, and peripheral blood lymphocytes. Levels of intracellular glutathione have been shown to increase by 16% in T-cell cultures at concentrations of 10- 100 [M (concentrations achievable with oral and intravenous supplementation of ALA). Increases in glutathione levels seen with ALA administration are not only from the reduction of oxidized glutathione (one of the functions of ALA) but also from the synthesis of glutathione."

Cilantro: Mobilizes toxic metals from the central nervous system and other tissues. A researcher named Dr. Yoshiaki Omura, using bioenergetic measures, discovered that some patients excreted more toxic metals after consuming a Chinese soup containing cilantro. Cilantro is the leafy part of a common herb whose seed, coriander, is a familiar culinary spice. Its active component is a mercaptan that can penetrate the blood brain barrier.

Dr. Andrew Hall Cutler says that Omura is right that cilantro contains some active principle that effectively binds (and releases) mercury and crosses the blood brain barrier but says that the pharmacology or kinetics of cilantro's active principle remains unknown. Dr. Hall has deterred people from using it for he has maintained that nobody has any clue how to use cilantro. This issue has been resolved by Drs. Greenburg and Georgiou and several others who use cilantro safely and effectively.

Cilantro stimulates the body's release of mercury and other heavy metals from the brain and CNS into other tissue. Cilantro's postulated mechanism of action is to act as a reducing agent changing the charge on the intracellular mercury to a neutral state allowing mercury to diffuse down its concentration gradient into connective tissue.

Cilantro has a health-supporting reputation that is high on the list of the healing spices. Cilantro has been well-researched and has been found to have many benefits including blood sugar lowering properties,[13] anti-inflammatory properties – contains flavonoids include quercitin, kaempferol, rhamnetin, and epigenin, free radical scavenger and prevents lipid peroxidation properties,[14] and is seen as antimicrobial due to being rich in volatile oils such as carvone, geraniol, limonene, borneol, camphor, elemol, and linalool. Research performed in Mexico and United States has isolated the compound dodecanal – which laboratory tests showed is twice as effective at killing Salmonella as the antibiotic gentamicin.[15]

Chlorella: Chlorella plays a particularly crucial role in systemic mercury elimination because the majority of mercury is rid through stool. Once the mercury burden is lowered from the intestines, mercury from other body tissues will more readily migrate into the intestines — where chlorella will effectively remove it. It is the fibrous material in chlorella that has been shown to bind with heavy metals and pesticides like PCBs that can accumulate in our bodies. Chlorella traps toxic metals in the GI tract and acts as an ion exchange resin.

Chlorella is a species of unicellular fresh water algae that has been shown to possess detoxifying properties enabling it to assist or support the human detoxification system. Chlorella algae contain phytochemicals that support detoxification while the cell walls function as an ion exchange resin to absorb and retain toxic metals which can then be excreted. Chlorella is a food-like all purpose mild detoxifier (not chelator) of heavy metals.

The detoxification capability of chlorella is due to its unique cell wall and the material associated with it. The cell walls of chlorella have been shown to have three layers of which the thicker middle layer contains cellulose microfibrils. Atkinson et al found a 14nm thick trilaminar layer outside the cell wall proper which was extremely resistant to breakage and thought to be composed of a polymerised carotene like material.

Laboratory studies showed that there were two active absorbing substances — sporopollenin (a naturally occurring carotene like polymer which is resistant to degradation) and the algae cell walls. The fibrous material augments healthy digestion and overall digestive track health.

Chlorella's cleansing action on the bowel and other elimination channels, as well as its protection of the liver, helps keep the blood clean. Clean blood assures that metabolic wastes are efficiently carried away from the tissues. Chlorella gets its name from the high amount of chlorophyll it possesses.

Chlorella contains more chlorophyll per gram than any other plant. Chlorella can speed up the rate of cleansing of the bowel, bloodstream and liver, by supplying plenty of chlorophyll. Chlorella and spirulina are used as nutrient-dense foods and sources of fine chemicals in their most natural forms. They have significant amounts of lipid, protein, chlorophyll, carotenoids, vitamins, minerals, and unique pigments. They may also have potent probiotic compounds that enhance health. (You will find three chapters in *Survival Medicine for the 21ˢᵗ Century* devoted to Spirulina as a pure medicine.) Both have extensive scientific research[16] that all indicate their value for a wide range of medical situations.

Mercury can also be bound to sulfhydryl groups in garlic or to sulfur in the form of MSM.

There are many approaches, substances, companies and doctors offering detoxification and chelation products. The great problem is finding science and in-depth studies to back up the many claims being made.[17] Currently, the only product that uses the golden triangle of chlorella, cilantro and ALA is Chelorex™ (*key-lor-ex*). It is also one of the most tested natural chelation products on the market.

Science Formulas, developer of Chelorex, performed hair, fecal and urine tests on 151 subjects and the results are impressive.[18] It fulfills the ideal of promoting excretion via multiple pathways in the body.

There are several natural chelation products that use only the cilantro and chlorella to extremely positive effect supporting the basic premises being put forth here. The addition of ALA brings in the leading work of Dr. Andrew Hall Cutler, who is one of the world's leading experts on mercury detoxification. His extensive and successful use of ALA has won him a large devoted audience.

In a large metal foundry in Russia, Dr. George Georgiou tested extensively many natural substances for their efficacy in removing heavy metals from the workers there and found chlorella, chlorella growth factor and cilantro so effective - when used "together" - that he introduced Heavy Metal Detox (HMD™) in 2005.[19] But the issues are not straight or clear cut as Dr. Georgiou explains, "During the three years that I have been researching the efficacy of certain natural substances for their heavy metal chelating effects, I have stumbled across a few surprises. For example, the literature was full of testimonials on how chlorella and cilantro are excellent chelators of heavy metals, so we tested both of these in carefully designed, double-blind, placebo controlled trials.

Let's take *chlorella vulgaris* as an example - when we tested this alone in pre-post provocation urine and feces tests using 3,000 mg daily, we found no difference between the pre and post tests. In other words, chlorella by itself was not eliminating any metals that could be detected by an ICP-MS at parts per billion levels of measurement."

Georgiou continues saying, "When we looked at cilantro we were even more surprised as there was a percentage *decrease* in the post-test compared to baseline. What this means was that the Cilantro not only was not eliminating metals but it was actually absorbing more metals than baseline levels. It is very probable that Cilantro, which is known as an intracellular chelator, takes metals from the interior of the cell and brings them out into the mesenchyme or extracellular space. As there is nothing to mop them up here, as the osmotic gradient increases, then you get a rush of metals from the extracellular environment into the intracellular environment.

Personally, I have seen patients who were given only cilantro by other practitioners get worse while on this protocol. Based upon the extensive double-blind placebo controlled trials that I have run with 350 people, I would strictly avoid using cilantro by itself with no other backup." This experience of Dr. Georgiou mirrors the warnings of Dr. Cutler about using cilantro alone or indiscriminately.

After 3 years of research, Dr. Georgiou settled on a different golden triangle of natural chelation using Chlorella Growth Factor, Coriander sativum leaf (cilantro) and an energized cell-decimated homaccord of *chlorella vulgaris*. Dr. Greenburg's choice of ALA plus the addition into his formula of selenium, magnesium, zinc, vitamin C, E, MSM (Methysulfonylmethane), NAC (N-Acetyl-L-Cysteine), L-Glutamine and Taurine turns Chelorex into a complete nutritionally driven chelation prescription effective for and tested for the removal of 16 toxic metals.

Much needs to be said about all of these other substances in Chelorex and we will devote a new book (*Survival in the 21st Century*, which is where this chapters material is sourced) on basic principles like the promotion of glutathione.

A major point about successful chelation, especially of mercury, is seen in the example of the importance of selenium. The detoxifying effect of selenium on *mercury* toxicity is due to a formation of a biologically inactive complex containing the elements in an equimolar ratio. The complex is unable to pass biological barriers, placenta and choroid plexus and is stored in the liver and the spleen, even in the brain in a non toxic form. Mercury binds with selenium and selenium is needed for glutathione production in the cells. Mercury-selenium-glutathione interactions are crucial to anyone who wants to understand the consequences of mercury exposure and how to combat or detoxify from its poisonous effects.

Both Chelorex and HMD may not be as "powerful" as synthetic chelators such as DMSA, DMPS, EDTA etc., in the sense they do not eliminate the more narrow band of metals these synthetic chelators were designed to remove as quickly as these others when applied intravenously, but neither Chelorex or HMD have the side effects of these synthetic compounds.

The main advantage of natural chelators over the synthetic ones is that they do not strip the body of essential minerals, which is common with synthetic chelators.

With the kinds of oral minerals that most doctors use today, which are not in forms that are easily absorbed by the body, this demineralization is critical. The natural chelation protocols that are suggested here go hand-in-hand with raising magnesium levels, using magnesium chloride and concentrated sea mineral formulas to supplement orally a broad range of healing minerals.

NAC (N-Acetyl Cysteine- oral) gives so much value for the money that the need for glutathione can be met for days with this oral product. NAC (N-acetyl-Cysteine) is proven more effective than IV glutathione for acute liver toxicity.
                                                    Dr. Garry Gordon

Synthetic chelators, especially DMSA and DMPS are actually quite toxic and one can hardly imagine using them for years. Dr. Jaquelyn McCandless reminds us, "Oral agents, especially DMSA, can encourage yeast overgrowth." When chelating people with a heavy metal burden, particularly when they are young children or very elderly, or have any chronic disease, it is best to mobilize and eliminate the metals gently, slower than faster, so that the body can reabsorb less and avoid flooding the body with toxic metals that cause further oxidative stress due to their free radical activity. Dr. Timothy Ray, an oriental medical doctor speaks elegantly about avoiding the healing crisis that synthetic chelators so often bring. He has a product similar to HMD called NDF and NDF Plus, which are also based on chlorella and cilantro.

MSM aids in detoxifying metals by
contributing sulfur to methionine and cysteine.[20]

Chelorex stands out in several powerful ways; the first being the broad array of metals it will eliminate, including uranium. We find this in the HMD product as well. Chelorex though contains ingredients that stimulate the human body's natural immune defenses including its own innate ability to remove heavy metals.[21] In other words, instead of further compromising the patient's body by adding more toxins, it adds nutrients and attractors that allow existing toxins to be flushed out.

Most importantly it gives a solid baseline of minerals that are absolutely essential for both short-term and long-term detoxification and chelation of toxic substances. The formula gives an important boost to the critical minerals supporting the use of a broad based sea mineral concentrate supplement, which is the best way to get one's minerals besides from foods.

The individual elements found in HMD and Chelorex
are all natural substances, which apart from their
chelating properties have many other health benefits.

Dr. George Georgiou was sponsored by a private company who invested about $1 million dollars in a double-blind, placebo controlled trial with 350 people, that has shown its ability to safely chelate a variety of metals with natural substances, that when combined, work as effectively as synthetic chelators. This resulted in the production of HMD.[22] Natural chelation is safe, non-invasive, affordable and available without prescription as it is considered food supplement.

Dr. Alan Greenberg invested a small fortune also and closely studied 151 people using hair, fecal and urine tests. Together these doctors have proven conclusively the power and safety of natural chelation. The bottom line is that we want to get the heavy metals out and do that while introducing no new harmful agents. Extensive studies and tests prove that this capability is now available.

There is a tremendous amount of anecdotal evidence with a host of chelation products and substances on the market but very little hard proof. It is far easier to make claims than to prove through heavy investments in studies.

Human exposure to heavy metals has risen dramatically in the last fifty years as a result of an exponential increase in the use of heavy metals in industrial processes and products. The need for detoxification and chelation is increasing considerably and yet, exaggerated healing crises or residual side-effects caused by detoxification could arise, caused by fundamental deficiencies in magnesium and other minerals.

According to Dr. Sherry Rogers, there is as much as a five hundred-fold variance in ability, from one individual to the next, with regard to detoxifying the same chemical. One of the key factors of this difference is each individual's magnesium level.

There are over two hundred published clinical studies[22] that document the need for magnesium and many examples of miraculous "cures" from the use of this common mineral. Yet you'd wonder if anyone is reading them. As one example, doctors for the advocacy group, Defeat Autism Now (DAN) underestimate autistic children's needs; recommending only 50 milligrams twice a day in oral form, even though children with gastrointestinal problems can absorb only small percentages through their intestines.

The entire autism community needs to be acutely aware that its present dependency on oral magnesium supplementation contributes to the less than effective results from chelation. A simple changeover to transdermal and topical magnesium supplementation approaches would yield far more effective results.

Dr. Leslie Fisher has treated over of 35,000 patients where mineral therapy was prescribed as the sole form of medication with good results. He has conducted research in the Department of Psychiatry at Austin Hospital, in Melbourne, Australia. Mineral therapy is the foundation that chelation treatments and protocols should to be built on. An over reliance on synthetic chelators is dangerous without appropriate mineral support therapies. In the case of autism spectrum disorders, oral magnesium supplement plans cannot be expected to alleviate magnesium deficiencies.

Sufficient magnesium levels lead to safer detoxification and chelation. In fact, it makes chelation possible, as Dr. Boyd Haley points out in the next section.

The autism community has been told by Dr. Andrew Hall Cutler that "mineral transport" defects are the biggest problem with heavy metal toxicity. The discovery that an important mineral like magnesium can be supplied through the transdermal route is significant for the autism community and everyone else. The last thing any parent or doctors treating autistic children want to do is to spend any more time using a magnesium supplementation modality that has shown itself to be ineffective.

References

1   Chelation therapy is a recognized treatment for heavy metal (such as lead) poisoning. EDTA, injected into the blood, will bind the metals and allow them to be removed from the body in the urine. (www.americanheart.org)

2   According to tests done in the US and Japan perfluorinated chemicals (PFCs) have crept into the blood of almost every living creature in the northern hemisphere. The US Environmental Protection Agency has begun an investigation to determine how a Teflon chemical has found its way into the blood of virtually every American, and polluted drinking water supplies. Perfluorooctanoic acid, a key ingredient in the making of Teflon non-stick coating for cookware can cause testicular, breast, liver and prostate cancers, as well as birth defects. And in 2004 the U.S. Food and Drug Administration has found perchlorate contamination in nearly all of the more than 200 samples of lettuce and milk it collected and tested nationwide. The federal government has not yet established a standard for the perchlorate, but the Environmental Protection Agency has adopted a provisional recommendation that contamination in drinking water not exceed a range of 4 to 18 parts per billion. Perchlorate, a chemical used in rocket fuel, munitions and fireworks, can affect thyroid gland functions and lead to developmental difficulties in children. The FDA found perchlorate in 217 of 232 samples of milk and lettuce in 15 states. In 104 samples of milk, the average concentration was 5.76 parts per billion. In 128 samples of lettuce, the average concentration was 10.49 parts per billion.

3   "Most adults in the Netherlands are exposed to approximately 2 picogrammes of toxic equivalents of dioxin-like substances per kilogramme of body weight per day. In general, it may be stated that in excess of 90% of human exposure to PCDDs, PCDFs and dioxin-like PCBs derives from the consumption of animal fats, of which 50% are contained in milk and milk products. Infants are exposed to these substances before birth as well as through the maternal milk," reported the Health Council of the Netherlands. See: www.borstvoeding.com/abon/bf_toxins.html

4   The Atlantis Mobile Laboratory, just tested in Bermuda, reported that 50 out of 70 newborns surveyed had dangerously high levels of mercury. Another study led by Mount Sinai School of Medicine in New York, through blood sampling, "found an average of 91 industrial compounds, pollutants, and other chemicals in the blood and urine of nine volunteers. The people tested did not work with chemicals on the job or live near an industrial facility. The body burden of a total of 167 chemicals found in the volunteers, showed 76, which cause cancer in humans or animals, 94, which are

toxic to the brain and nervous system, and 79 that cause birth defects or abnormal development. The dangers of exposure to these chemicals in combination have never been studied. Body Burden. National Report on Human Exposure to Environmental Chemicals. For more info, see: www.oztoxics.org/cmwg/body%20burden/international. html

5   Heavy metal poisoning has become an increasingly major health problem, especially since the industrial revolution. Heavy metals are in the water we drink, the foods we eat, the air we breathe, our daily household cleaners, our cookware and our other daily tools. A heavy metal has a density at least 5 times that of water and cannot be metabolized by the body, therefore accumulating in the body. Heavy metal toxicity can cause our mental functions, energy, nervous system, kidneys, lungs and other organ functions to decline.

6   "Data from the CDC tells us that children are carrying around more phthalates and certain pesticides in their bodies than adults and that woman have more mercury and some other toxic chemicals in their bodies than men. This is very disturbing because children and babies *in utero* have some of the highest risks of adverse health impacts," said Charlotte Brody, RN, executive director of Health Care Without Harm.

7   The World Wildlife Fund (WWF) visited 13 locations in England, Northern Ireland, Scotland and Wales in the summer of 2003 and took blood samples from 155 volunteers. Lancaster University analyzed the samples for 78 chemicals. 12 organochlorine pesticides (including DDT and lindane), 45 PCB congeners and 21 polybrominated diphenyl ethers (PBDE) flame retardants, including those found in the commercially traded penta-, octa- and deca-BDEs.

Their FINDINGS:

• Every person tested is contaminated by a cocktail of known highly toxic chemicals which were banned from use in the UK during the 1970s and which continue to pose unknown health risks.

• We found 70 (90 per cent) of the 78 chemicals we looked for in the survey. The highest number of chemicals found in any one person was 49, nearly two thirds (63 per cent) of the chemicals looked for.

• Every person is contaminated by chemicals from each group: organochlorine pesticides, PCBs and PBDEs (flame retardants).

• The highest concentration of any chemical found was 2,557 ng/g (ng/g. parts per billion) of the DDT metabolite pp.-DDE. The use of DDT was banned in the UK more than 20 years ago.

• The most frequently detected chemicals were PCB congeners 99 and 118 and the DDT metabolite pp.-DDE, which were detected in all but one of the 155 volunteers.

• Ten chemicals were found in more than 95 per cent of volunteers

8   www.gordonresearch.com/answers/chlorella_and_cilantro.html

9   The earliest types of chelation involved synthetic agents such as BAL, penicillamin e and EDTA administered intravenously for acute toxic metal poisoning. Subsequently, DMSA and DMPS were utilized, first intravenously and later orally and now even transdermally. Chelation therapy provides a relatively safe, effective, and inexpensive alternative to the drugs and surgery often used for circulatory disorders such as coronary heart disease, carotid (neck artery) stenosi (blockage), and leg artery stenosis (blockage). Chelation is a process by which toxic substances in the body, particularly heavy metals can be excreted safely. However, numerous negative side effects are associated with each of these chelators (with the exception of EDTA which is 'relatively' safe) including allergic reactions involving the skin and mucous membranes (itching, exanthema or rash), as well as occasional cases of Stevens-Johnson Syndrome or erythema exudative multiforme.(11). Other side effects include nausea, headache, muscle aching, changes in taste, severe malaise, dizziness, numbness, insomnia, diarrhea, weight loss, extreme fatigue, leg cramps, cardiac arrhythmia, liver and kidney damage, abdominal pain, anxiety, severe restlessness, mental changes, tremors, inability to concentrate, poor memory, impaired equilibrium, chemical sensitivities and tinnitus. (12). Studies have shown up to 30% of patients have severe negative side effects as a result of these synthetic chelating agents, which may develop after a single dose. For warnings against DMPS all one has to do is go to the DMPS Backfire website.

10   R. A. Goyer, M. G. Cherian, M. M. Jones, and J. R. Reigart. Role of Chelating Agents for Prevention, Intervention, and Treatment of Exposures to Toxic Metals. *Environmental Health Perspectives* Volume 103, Number 11, November 1995

11   Jones MM, Cherian MG. The search for chelate antagonists for chronic cadmium intoxication. *Toxicology* 1990;62:1-25.

12   Korkina, L.G., et al. Antioxidant therapy in children affected by irradiation from the Chernobyl nuclear accident. *Biochem. Soc. Trans.*, Vol. 21, 1993, p. 314S

13   Gray AM, Flatt PR. Insulin-releasing and insulin-like activity of the traditional anti- diabetic plant Coriandrum sativum (coriander). *Br J Nutr* 1999 Mar;81(3):203-9.

14   Chithra V, Leelamma S. Coriandrum sativum changes the levels of lipid

peroxides and activity of antioxidant enzymes in experimental animals. *Indian J Biochem Biophys* 1999 Feb;36(1):59-61.

15   Delaquis PJ, Stanich K, Girard B et al. Antimicrobial activity of individual and mixed fractions of dill, cilantro, coriander and eucalyptus essential oils. *Int J Food Microbiol.* 2002 Mar 25;74(1-2):101-9.

16   http://www.mercola.com/chlorella/research.htm

17   In the area of chelation we have many companies making claims. For example a product released by Dr. Eliaz of modified citrus pectin and dietary alginate claims, "Not only that, *Pectasol*® is the only chelator of its kind that's been shown to eliminate toxic heavy metals and other dangerous substances on humans. A recent pilot clinical trial – in which patients were tested for 20 different toxic elements and minerals – showed dramatic results." The data of the (Pre clinical results) shows increased urinary excretion over 6 days of 3 metals (arsenic, mercury & cadmium) not 20 as advertised and does it with a small group over a short period of time. Comparing this study to Science Formulas' Chelorex 4yr study or to Dr. George Georgiou's studies in Russia is just not possible. There are many products available and the choices are bewildering. Words are easily said and written but when it comes to removing heavy metals it is very important to know what a product can do.

18

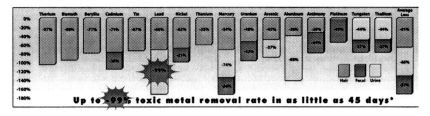

19   http://mercuryexposure.org/index.php?article_id=648

20   **MSM (methylsulfonylmethane):** Enhances permeability of cell membranes and is believed to enhance detoxification by increasing the permeability of cell walls. MSM is a naturally occurring sulfur containing molecule found in fruits, vegetables, seafood and meat. It is present in the body and humans excrete from 4-11 mg. daily in urine. Research suggests that it is required for the body to preserve normal function and structure. Food processing destroys the MSM normally present in food.

21   Cysteine and cystine are closely related. One cystine molecule is composed of two bonded cysteine molecules and each can convert to the other as required. Both amino acids contain sulfur (via free sulfhydryl groups) which makes them powerful

antioxidants. **The acetylated form of cysteine is N-acetylcysteine (NAC)** and contains a bonded acetyl group. In this form, NAC is more easily absorbed, more stable, and safer to use than cysteine on its own, which can be neurotoxic in very high doses. NAC is effective at promoting glutathione synthesis. This amino acid "combo" incorporates cysteine, glutamic acid, and glycine and has powerful antioxidant and immune stimulating properties. Some studies have shown that supplementing with NAC yields higher glutathione levels than supplementing with cysteine or glutathione directly. Cysteine is found in alpha-keratin, the main protein of fingernails, toenails, skin, and hair.

22   See: http://mgwater.com

9

||||||||||||||||||||||||||||||||||||||||||||||||||||||||||||||||||||||

Magnesium and ATP

Magnesium is critical for all of the energetics of the cells because it is absolutely required that it be bound (chelated) by ATP (adenosine triphosphate), which is the central high energy compound of the body.

ATP that is not bound to magnesium cannot create the energy normally used by specific enzymes that the body uses to make protein, DNA, RNA, transport sodium or potassium or calcium in and out of cells, nor to phosphorylate proteins in response to hormone signals, etc. In fact, ATP without enough magnesium is non-functional and leads to cell death.

Bound magnesium holds the triphosphate in the correct stereochemical position so that it can interact with ATP using enzymes. In addition, magnesium polarizes the phosphate backbone so that the "backside of the phosphorous" is more positive and susceptible to attack by nucleophilic agents such as hydroxide ions or other negatively charged compounds.

"Bottom line, magnesium, at critical concentrations is essential to life," says Dr. Boyd Haley who asserts strongly that, "All detoxification mechanisms have as the basis of the energy required to remove a toxicant the need for magnesium and ATP to drive the process. There is nothing done in the body that does not use energy and without magnesium. This energy can neither be made nor used."

What do the following disorders have in common:

PMS? Sinus problem? Arthritis? Chronic Fatigue? Fibromyalgia? Lupus? Psoriasis? Painful and enlarged breasts?

They are all chronic conditions that are fully or partially due to a backup of the body's lymph system. These conditions were created over a long period of time as the body's immune system had been gradually weakened.

The cause of this weakening is a slowdown of the metabolism which is often goes hand-in-hand with magnesium deficiency. A slow metabolism is the profound and most significant effect of an overabundance of waste products within the tissues. In a healthy body, the lymph fluid carries waste away. An interference of this process is often the beginning of chronic conditions such as mentioned above.

Mercury drastically increases the excretion of magnesium and calcium from the kidneys.[1] Both mercury itself and the drugs used to chelate mercury have a strong impact on mineral levels. Between the dramatically diminishing mineral content in foods and the highly toxic world we live in, the need for magnesium has never been greater.

A magnesium ion is an atom that is missing two electrons, which makes it search to attach to something that will replace its missing electrons so it is actively and directly involved in diminishing heavy metal toxicity. Magnesium appears to be a competitive inhibitor of lead and cadmium.[2] An increased level of magnesium has been shown to eliminate lead and cadmium through the urine[3] and has also been reported to reduce the toxic effects of aluminum.[4]

Magnesium ions constitute the physiologically active magnesium in the body; they are not attached to other substances and are free to join in biochemical body processes.[5] This is one basic reason magnesium helps to detoxify toxic chemicals and helps eliminate heavy metals from the body. Another reason would be the part it plays in glutathione production but undoubtedly, as Dr. Haley indicates, the Mg-ATP provides the crucial energy to remove each toxicant. Thus Dr. Dean is certainly justified in saying, "Symptoms of chemical toxicity can be completely or partially produced by magnesium deficiency."

Transdermal magnesium chloride therapy (the only way one can dependably increase magnesium levels besides intravenous application) is crucial in any kind of detoxification or chelation program. Deficiencies of magnesium result in a deficiency in vitamin B_6 because it is involved in vitamin metabolism. If the body doesn't get enough magnesium, it cannot make or utilize protein.

Magnesium deficiency creates other problems, such as protein deficiencies. Magnesium also activates vitamins C and E. Therefore, if there isn't enough magnesium in the body, vitamin C and E consumed in foods will likewise be unused. Without enough magnesium, DMPS and DMSA — the most commonly used synthetic chelators — will hurt more than they will help.

10

||||||||||||||||||||||||||||||||||||||||||||||||||||||||||||||||||||

Magnesium and Diabetes

Magnesium for diabetics is critical. A Gallup survey in 1995 of five hundred adults with diabetes reported that 83% of those with diabetes are consuming insufficient magnesium from food, with many by significant margins.[1] At least 25% of diabetics have hypomagnesemia[2] and this is likely an underestimate.

One group has recently suggested that the effects of reduced glutathione on glucose metabolism may be mediated, at least in part, by intracellular magnesium levels.[3]

Dr. Carolyn Dean indicates that magnesium deficiency may be an independent predictor of diabetes and that diabetics need more, and are prone to *losing* more magnesium than most people.

Magnesium is necessary for the production, function and transport of insulin. Magnesium deficiency is associated with insulin resistance and increased platelet reactivity. According to Dr.

Jerry L. Nadler, "The link between diabetes mellitus and magnesium deficiency is well known. A growing body of evidence suggests that magnesium plays a pivotal role in reducing cardiovascular risks and its reduced presence may be involved in the pathogenesis of diabetes itself.

"While the benefits of oral magnesium supplementation on glycemic control have yet to be demonstrated in patients, magnesium supplementation has been shown to improve insulin sensitivity. Based on current knowledge, clinicians have good reason to believe that magnesium repletion may play a role in delaying, if not eliminating, Type II diabetes onset and potentially in warding off its devastating complications — cardiovascular disease, retinopathy, and nephropathy."

The mechanism of hypomagnesemia in diabetic patients still remains unsolved. However, there is enough evidence to suggest that magnesium levels drop in the course of recovery from ketoacidosis, during insulin therapy[4] or with severe retinopathy[5] or proteinuria.[6]

Diabetic patients, especially those with poor glucose control, can also develop hypomagnesemia from a glucose-induced osmotic diuresis.

Insulin resistance and magnesium depletion may result in a vicious cycle of worsening insulin resistance and decrease in intracellular magnesium (Mg_2+) which may limit its effectiveness in vital cellular processes.

Diabetic ketoacidosis (DKA)[7] is a state of inadequate insulin levels resulting in high blood sugar and accumulation of organic acids and ketones in the blood. Increased blood acids (ketoacidosis) can be an acute complication of diabetes. It occurs when your muscle cells become so starved for energy that your body takes emergency measures and breaks down fat, a process that forms acids known as ketones.[8]

Hyperglycemia initially causes the movement of water out of cells, with subsequent intracellular dehydration, extracellular fluid expansion and hyponatremia (sodium loss). It also leads to a diuresis in which water losses exceed sodium chloride losses. It is believed that magnesium is also lost by osmotic action. Urinary losses then lead to progressive dehydration and volume depletion, which causes diminished urine flow and greater retention of glucose in plasma. The net result of these alterations is hyperglycemia with metabolic acidosis.[9]

Proteinuria is protein in the urine, caused by damaged kidneys and a declining ability of the kidneys to protect the body from protein loss. This is frequently seen in longstanding diabetes, hypertension, as well as other chronic renal conditions. In the United States, diabetes is the leading cause of end-stage renal disease (ESRD), the result of chronic kidney disease. In both Type I and Type II diabetes, the first sign of deteriorating kidney function is the presence of small amounts of albumin in the urine, a condition called *microalbuminuria*. As kidney function declines, the amount of albumin in the urine increases, and microalbuminuria becomes full-fledged proteinuria.

Lower serum magnesium levels are associated with more rapid decline of renal function. During insulin treatment, neither magnesium nor potassium can be metabolized properly, so these essential minerals must be replenished.

Severe symptomatic hypermagnesemia is relatively rare, but high levels of magnesium can develop in people, most commonly those with renal insufficiency or renal failure.[10] Kidney disease, rather than diet, is the usual cause of magnesium overload, because the kidneys lose the ability to remove excess magnesium.

Diabetic Children and Magnesium

On Monday, November 7, 2005, *The Associated Press* reported that "About two million U.S. children ages 12 to 19 have a pre-diabetic condition linked to obesity and inactivity that puts them at risk for full-blown diabetes and cardiovascular problems, government data suggest." 1 in 14 boys and girls in a nationally representative sample had the condition. Among the overweight adolescents, it was 1 in 6. The study in question appears in the November (2005) edition of *Pediatrics*. It is based on data involving 915 youngsters who participated in a 1999-2000 national health survey.[11]

A genuine disaster in autism is unfolding at a rate that is now affecting 1 child in approximately 166. However, if all children with severe learning disabilities are counted, the ratio skyrockets to 1 in 6.

What is happening to the nation's children? The CDC cannot shed any light on the subject because it has its head stuck in the bird flu sand. The FDA offers nothing of real significance because it is too busy protecting the role of pharmaceutical companies' "franchise" in the treatment of disease. Unfortunately, their proprietary solutions most often involve attempts to cure by way of unnatural toxification, and not by the natural nourishing and strengthening the body. While patients in all age groups are affected by this strategy, children are most vulnerable.

Dr. Teresa A. Hillier has reported, "Diabetes increased the risk of heart attack and stroke in both age groups, but the increased risk was much larger in younger people. People who had been diagnosed before age 45 were 14 times more likely to have a heart attack and 30 times more likely to have a stroke than their non-diabetic peers. In contrast, older people with diabetes were four times more likely than their peers to have a heart attack and three times more likely to have a stroke." Hillier and a colleague, Kathryn L. Pedula, based their findings on a study of nearly 8,000 people who were newly diagnosed with Type II diabetes.[12]

While Type II diabetes used to be primarily a problem of middle and old age, new cases of the illness among people 30 to 39 have risen 70% in the last decade.

What is happening to our children by the current state of medical affairs is a disaster and no words can express the pain and agony that millions of parents are facing in the United States alone. Some day the medical authorities will be held responsible for their failure to address these issues. In the meantime, it is incumbent upon each individual and family member to inform one's self.

In the area of autism spectrum disorders, the government just does not want to admit that hundreds of thousands of children have been damaged by vaccines laden with mercury. In actuality, *poisoned* would be the more appropriate word, but no one at the FDA or CDC will publicly acknowledge anything about the dangers of low level toxicity because such disclosure in those areas would raise vexing questions, charges, accusations of conspiracy and guilt, and even criminal prosecution.

> *Diabetes has risen by over 14% in the last*
> *two years. The CDC estimates that 20.8 million*
> *Americans — 7% of the U.S. population —*
> *have diabetes, up from 18.2 million in 2003.*[13]
>
> Centers for Disease Control

Is a lack of magnesium related to Type II diabetes in obese children? Dr. Huerta and colleagues say yes in their study titled *Magnesium deficiency is associated with insulin resistance in obese children.*[14]

Insulin resistance occurs when the body does not use insulin, a protein made by the pancreas, to turn glucose into energy. Children who are obese (seriously overweight) are more likely to have insulin resistance. This might be because they have low magnesium levels in their blood.

Dr. Huerta's study was done to see if obese children get enough magnesium in their diets and if a lack of magnesium can cause insulin resistance and eventually Type II diabetes. This is the first study linking low magnesium levels to insulin resistance in obese children.

Researchers found that 55% of obese children did not get enough magnesium from the foods they ate, compared with only 27% of lean children. Obese children had much lower magnesium levels in their blood than lean children. Children with lower magnesium levels had a higher insulin resistance.

The results of the diet survey showed that obese children processed 14.4% less magnesium from the foods they ate than lean children, even though obese and lean children ate about the same number of calories per day. Obese children eat more calories from fatty foods than lean children. In addition to not eating enough foods rich in magnesium, obese children seem to be less proficient in using magnesium from the foods they eat. Extra body fat appears to prevent the body's cells from using magnesium to break down carbohydrates.

When it comes to diabetes there is no lack of information pointing to magnesium deficiency and chemical poisoning converging on the young, but rather than acknowledging treatment options which may not require a pharmaceutical response, medical authorities have done little to address nutritional and magnesium deficiencies while instead, warning parents of chemical dangers.

The United States government seems to be involved in a huge cover up of medical and pharmaceutical wrong doings, which are supported by the insurance industry (which only pays for costly prescription drugs and surgical procedures) and the media (which often casts doubts and aspersions against "alternative" or natural treatment modalities), and will just keep on letting things slide as hundreds of thousands of new kids *needlessly* get sick each year.

Magnesium and Diabetic Neuropathy

Diabetes is commonly thought to have no cure. It is progressive and often fatal, and while the patient lives, the mass of medical complications it sets off can attack every major organ. Though public health officials acknowledge that their ability to slow the disease is limited, and though doctors fear a huge wave of new cases will overwhelm public health systems, "Public health authorities around the country have all but ignored chronic illnesses like diabetes, focusing instead on communicable diseases, which kill far fewer people," according to the New York Times.

Hospitals around New York City are full of diabetic patients and on any given day, nearly half the patients are there for some trouble precipitated by the disease.[15]

Type II diabetes is being declared
an epidemic in New York City.

With 1 in 3 children born in the United States five years ago expected to become diabetic in their lifetimes, a close look at its surge in New York City offers a disturbing glimpse of where the city and the rest of the world is headed. Diabetes has swept through families, entire neighborhoods in the Bronx and broad slices of Brooklyn. While the ranks of American diabetics have exploded by an extremely painful 80% in the last decade, New York has seen a devastating explosion of *140%*.

The best that the news media can do in their mission to "inform the public," is to *not* question, challenge, or otherwise investigate the notion that diabetes is incurable, but instead, to have viewers brace themselves for more, and to learn to "live with it." As long as we still have the ability to procreate (to replace the hundreds of millions who needlessly suffer and die prematurely), this would ensure a rosy future for drug manufacturers, and a bleak outlook for our nation's health.

Type II Diabetes is sweeping so rapidly through
America we need not waste time giving
children bicycles. Just roll them a wheelchair.
Boston Globe[16]

This chapter on diabetic neuropathy introduces a much needed medical intervention for the prevention and treatment of diabetes and the many complications that come from it, but as the New York Times admits, "In the Treatment of Diabetes, Success Often Does Not Pay."

"It's almost as though the system encourages people to get sick and then people get paid to treat them," said Dr. Matthew E. Fink, a former president of Beth Israel Medical Center in Manhattan. *The Times* bemoans, "A medical system so focused on acute illnesses that it is struggling to respond to diabetes, a chronic disease that looms as the largest health crisis facing the city."[17] Something is wrong with the way the medical establishment is dealing with diabetes and that starts with its refusal to look honestly at what is causing the disease.

> *Commercials tell children that junk food is good food*
> *- the latest message from an industry that spends*
> *$10 billion a year marketing to children.*
> New York Times

Medical science has discovered how sensitive the insulin receptor sites are to chemical poisoning. Metals such as cadmium[18], mercury[19], arsenic, lead, fluoride[20] and possibly aluminum may play a role in the actual destruction of beta cells through stimulating an auto-immune reaction to them after they have bonded to these cells in the pancreas. Food is not considered junk just because of high fat or sugar content, *there is a long list of poisonous chemicals used by the food industry that are striking mothers and daughters, fathers and sons, sisters and brothers, down.* There are many serious nutritional deficiencies in today's food that diminish the body's capacity to deal safely with these chemicals and heavy metals — with magnesium and selenium deficiencies at the top of the list.

For instance, according to Dr. Ellen Silbergeld, a researcher from the Johns Hopkins School of Public Health, the poultry industry's practice of using arsenic compounds in its feed is something that has not been studied: "It's an issue everybody is trying to pretend doesn't exist."[21]

Arsenic exposure is a risk factor[22] for diabetes mellitus.[23] Inorganic arsenic is considered one of the prominent environmental causes of cancer mortality in the world. Chicken consumption may contribute significant amounts of arsenic to total arsenic exposure of the U.S. population according to the *Journal Environmental Health Perspectives*.[24]

"Arsenic acts as a growth stimulant in chickens—develops the meat faster—and since then, the poultry industry has gone wild using this ingredient," says Donald Herman, a Mississippi agricultural consultant and former Environmental Protection Agency researcher who has studied this use of arsenic for a decade. Doctors also are on notice that many drugs have toxic effects that can participate as well in destroying insulin creation and cell receptivity to it.

Wistar rats were made diabetic
with a single injection of Alloxan.[25]

Another example is Alloxan. Studies show that Alloxan, the chemical that makes white flour look "clean and beautiful" destroys the beta cells of the pancreas.[26] Scientists have known of the alloxan-diabetes connection for years yet there seems to be a conspiracy in efforts to maintain the appearance of integrity in the

FDA, which allows dangerous chemicals that can cause diabetes to be used in drugs and food.

Consumer Reports recently observed, "A growing body of research shows that pesticides and other contaminants are more prevalent in the foods we eat, in our bodies, and in the environment than we thought,"[27] all confirming the chemical nightmare in progress.

We are just beginning to hear that a massive mistake has been made with genetically modified foods, which can only fan those diabetic winds.[28] Dr. Alpad Pusztai, for instance, has already shown that genetically-manipulated foods can, when fed to animals in reasonable amounts, cause very gradual organ and immune system damage.[29]

Another study, carried out by Dr Irina Ermakova at the Institute of Higher Nervous Activity and Neurophysiology, at the Russian Academy of Sciences, found that more than half of the offspring of rats fed on modified soya died in the first three weeks of life, six times as many as those born to mothers with normal diets.[30]

Dr. Manuela Malatesta and colleagues in the Universities of Pavia and Urbino in Italy, showed that mice fed on GM soya experienced a slowdown in cellular metabolism and modifications to liver and pancreas.[31] Researchers are reviving fears that GM food damages human health and certainly would not be indicated for children or people with diabetes.

Diabetic Neuropathy

Diabetic neuropathy, a complication of both Type I and Type II diabetes, is probably the most common complication of the disease.[32] Studies suggest that up to 50% of people with diabetes are affected to some degree. Diabetic neuropathy is a nerve disorder caused by diabetes. The two main classifications of neuropathy are *peripheral neuropathy*, affecting the extremities, arms, legs, hands and feet, and *autonomic neuropathy*, affecting the organ systems, mainly affecting the nerves of the digestive, cardiovascular systems, urinary tract and sexual organs.

Symptoms of peripheral nerve damage (neuropathy) are basically weakness, usually in the arms and hands or legs and feet, often with pain burning, tingling, or other abnormal sensations. Numbness or decreased sensation, difficulty walking and difficulty using the arms and hands or legs and feet are all common. Peripheral sensory neuropathy can initiate physiologic events that lead to distal extremity ulceration and eventual amputation.

Nerve damage caused by diabetes can also lead to problems with internal organs such as the digestive tract, heart, and sexual organs, causing indigestion, diarrhea or constipation, dizziness, bladder infections, and impotence.[33] Diabetic neuropathy is a major cause of impotence in diabetic men.[34] Autonomic neuropathies are believed to be implicated in "silent heart attacks" of diabetes, where the full symptoms of myocardial infarction are not felt by the person.

In some cases, neuropathy can flare up suddenly, causing weakness and weight loss. Neuropathy may cause both pain and insensitivity to pain in the same person. Often, symptoms are slight at first, and since most nerve damage occurs over a period of years, mild cases may go unnoticed for a long time. In some people, mainly those afflicted by focal neuropathy, the onset of pain may be sudden and severe.

A major risk factor of this condition is the
level and duration of elevated blood glucose.

Scientists do not know what causes diabetic neuropathy, but several factors are likely to contribute to the disorder. High blood glucose, a condition associated with diabetes, causes chemical changes in nerves. These changes impair the nerve's ability to transmit signals. High blood glucose also damages blood vessels that carry oxygen and nutrients to the nerves. Keeping blood sugar levels as close to the normal range as possible slows the onset and progression of nerve disease caused by diabetes.[35]

An increase in oxidative stress also may result in neuropathy. Hyperglycemia can increase intracellular sorbitol and fructose levels within neuronal tissue, which may lead to the production of harmful free radicals and an alteration of neuronal function.

It is possible to reduce amputation
rates by between 49% and 85%.[36]

Recently, researchers have focused on the effects of excessive glucose metabolism on the amount of nitric oxide in nerves. Nitric oxide dilates blood vessels. In a person with diabetes, low levels of nitric oxide may lead to constriction of blood vessels supplying the nerve, contributing to nerve damage. Scientists also know that high glucose levels affect many metabolic pathways in the nerves.

While the medical and the research industry is focused on the fact that hyperglycemia is the primary cause of peripheral neuropathy, little attention is paid to the effects of hypoglycemia,[37] or low blood sugars. This seems to be a far too ignored piece of the puzzle of diabetes, worsening as diabetics are pushed to achieve lower and lower blood sugars, with faster acting synthetic insulin. Research shows us that too rapid lowering of blood sugars from a longstanding hyperglycemic state,[38] and that episodes of sustained hypoglycemia can and do cause neuropathy.[39]

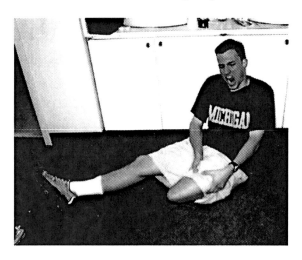

From 1993 to 1995, about 67,000 amputations were performed each year among people with diabetes. In 2002, about 82,000 non-traumatic lower-limb amputations were performed in people with diabetes.[40] The direct cost of an amputation associated with diabetes is estimated to be between US$30,000 and US$60,000. The estimated cost for three years of subsequent care ranges from US$43,000 to US$63,000 — mainly due to the increased need for home care and social services.

It has been forced upon me that diabetic
gangrene is not heaven sent, but earth born.
E.P. Joslin, 1934

Early in the last century, soon after the discovery of insulin, Joslin made the important observation noted above: he stated that it was not inevitable that a certain proportion of the diabetic population would develop foot ulceration or gangrene. He concluded that it was something to do with the way that we as health care professionals look after our patients, or the way that patients look after themselves, which results in conditions collectively referred to as 'the diabetic foot.'[41]

The medical establishment likes to think that great steps have been taken in diabetes management[42] but in the area of the diabetic foot, little impact has been made in the depressing statistics for rates of amputations and foot ulcers. Foot ulcers develop in approximately 15% of patients with diabetes, and foot disorders are a leading cause of hospitalization among such patients. 85% of lower-limb amputations in patients with diabetes are preceded by foot ulceration, suggesting that prevention and appropriate management of foot lesions are of paramount importance.[43]

Diabetic foot ulcer. Ulceration is caused by several factors acting together, but particularly by neuropathy.

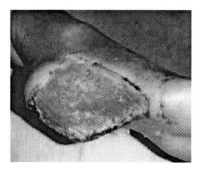

*Every 30 seconds a leg is still lost because
of diabetes somewhere in the world.*[44]

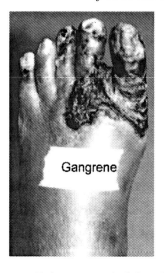

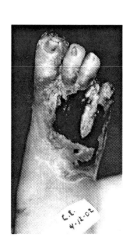

It is very painful to look at these pictures and imagine ourselves or a loved one with foot ulcers, gangrene, and eventual foot
amputation. One can get indignant knowing that just a little bit of
applied medical intelligence could avoid much of this.

A recent analysis showed that people with higher dietary intakes of magnesium (through consumption of whole grains, nuts,
and green leafy vegetables) had a decreased risk of Type II diabetes.[45] Magnesium has potentially beneficial effects at several key
steps of glucose and insulin metabolism. In animal studies, dietary

magnesium supplementation can prevent fructose-induced insulin resistance and elevations of blood pressure in rats. [46]

Magnesium and Diabetic Neuropathy

Magnesium is known to be necessary for nerve conduction; deficiency is known to cause peripheral neuropathy symptoms and studies suggest that a deficiency in magnesium may worsen blood glucose control in Type II diabetes. Scientists believe that a deficiency of magnesium interrupts insulin secretion in the pancreas and increases insulin resistance in the body's tissues.

An abstract from *Disorders of Magnesium Metabolism*[47] concludes, "Magnesium depletion is more common than previously thought. It seems to be especially prevalent in patients with *diabetes mellitus*. It is usually caused by losses from the kidney or gastrointestinal tract. A patient with magnesium depletion may present with neuromuscular symptoms, hypokalemia, hypocalcemia, or cardiovascular complication. Physicians should maintain a high index of suspicion for magnesium depletion in patients at high risk and should implement therapy early."

Diabetic neuropathy and other complications are most likely to be worse as a result of concurrent magnesium deficiency. Magnesium is known to be deficient in over 68% of the US population, and more so in diabetics who waste magnesium more than others when blood sugars are out of control.

Up to 80% of Type II diabetics have a magnesium deficiency.[48] Children labeled "pre-diabetic" — now *forty-one million* — are in great need of magnesium, which has been linked to preventing the development of Type II diabetes.[49]

In a series of papers, Dr. L. M. Resnick has shown in the test tube that an increase in glucose in the fluid leads to the release and/or displacement, of magnesium from the red blood cells, thus in the body hyperglycemia, high blood sugar, will cause a total body magnesium deficiency.[50]

A more recent study shows us that "Serum magnesium depletion is present and shows a strong relationship with foot ulcers in subjects with Type II diabetes and foot ulcers, a relationship not previously reported." Hypomagnesemia is associated with the development of neuropathy and abnormal platelet activity, both of which are risk factors for the progression of ulcers of the feet.[51]

Thus we can expect to find that magnesium can be used to prevent and treat both diabetes and the complications that come from it including severe peripheral neuropathy.

Dr. S. E. Browne makes a strong case for intravenous magnesium treatment of arterial disease and has used magnesium sulfate in his general practice for over three decades. "Magnesium sulfate ($MgSO_4$) in a 50% solution was injected initially intramuscularly and later intravenously into patients with peripheral vascular disease (including gangrene, claudication, leg ulcers and thrombophlebitis), angina, acute myocardial infarction (AMI), non-haemorrhagic cerebral vascular disease and congestive cardiac failure. A powerful vasodilator effect with marked flushing was noted after intravenous (IV) injection of 4-12 mmol of magnesium and excellent therapeutic results were noted in all forms of arterial disease."[52]

Dr. Herbert Mansmann Jr., Director of the Magenesium Research Lab,[53] who is a diabetic with congenital magnesium deficiency and severe peripheral neuropathy, shares that he was able to *reverse the neuropathy and nerve degeneration with a year of using oral magnesium preparations at very high doses.* "For example it took me 6 tabs of each of the following every 4 hours, Maginex, MgOxide, Mag-Tab SR and Magonate to get in positive magnesium balance. I tell people this not to scare them, but to illustrate how much I needed to saturate myself. Most will only need 10% of this amount. I was doing an experiment on myself to see if it helped my diabetic neuropathy. It worked so I did it for one year, and I have had *significant nerve regeneration.* I could never have been able to do this with $MgSO_4$ baths (Epsom Salt), since I could not get into and out of a bath tub."[54]

"I was saturated at about 3 grams of elemental magnesium per day, but went to 20 grams for over a year. I now take 5 grams, and stools are semi-formed, and the surrounding water is clear, 3-4 per day. Magnesium is very safe, since the gut absorption is regulated by serum magnesium levels, and then the magnesium stays in the gut and results in varying degrees of diarrhea. Then the dose is too high. Want soft semi-formed stools. Mine, while on high dosages of magnesium were liquid every 2-4 hours for 2 years, the electrolytes every month were normal, but for low potassium, part of my urinary magnesium wasting, both."

Dr. Mansmann concludes, "I have had diabetic neuropathy (DN) for over 10 years. The most significant symptom is my neuropathic pain of burning feet, called erythromelalgia (EM). With the aid of magnesium, I can completely suppress the symptom, but if my blood glucose level is acutely elevated, because of a dietary indiscretion, the pain flares in spite of an apparent adequate dose of magnesium. It goes away with extra magnesium gluconate (Magonate) in an hour or so in either case. Without the magnesium it will last for six plus hours, even though the blood glucose level is normal in about two hours."

"It is my belief that every one with diabetes should be taking magnesium supplementation to the point of one's Maximum Tolerated Dose, which is until one has soft semi-formed stools. In addition, anyone with neuropathy, without a known cause, must be adequately evaluated for diabetes and especially those with poorly, slowly, healing foot sores of any kind. Since the use of magnesium is safe I see no reason that this should not be "the standard of care".[55]

Conclusion

Prolonged use of Magnesium will prevent
chronic complications from diabetes.[56]

"The current 'party line' on this subject is not universally accepted, but many of us believe the establishment is too conservative and will some day change. While admitting its importance, for some unknown reason they remain reluctant to recommend magnesium supplements. They just do not know how poor the American diet is in magnesium and the frequency of magnesium deficiency" says Dr. Mansmann.[57]

Poorly controlled diabetes increases
loss of magnesium in urine.

It would be prudent for physicians who treat diabetic patients to consider magnesium deficiency as a probable contributing factor in many diabetic complications, as a main factor in exacerbation of the disease itself, and replenishment as a normal and customary first response.

The transdermal repletion of the patient's magnesium levels with magnesium chloride mineral therapy[58] is the ideal way of administering magnesium in medically therapeutic doses. Such treatments will, in all likelihood, help avoid and/or ameliorate complications such as diabetic peripheral neuropathy, arrhythmias, hypertension, and sudden cardiac death and will even improve the course of the diabetic condition in general.[59]

Once doctors, primary health care providers and the public are made aware of the role of magnesium in diabetes there will be no excuse to not increase public magnesium consumption, which can even be added to water supplies[60] instead of poisonous fluoride[61] and dangerous statins[62,63,64] which are also known to cause peripheral neuropathy with long term use.

During a stroke or heart attack it would be cruel, medically incompetent and life threatening to not use magnesium chloride or magnesium sulfate immediately. The same kind of treatment that saves lives in dramatic life threatening situations is urgently needed in the treatment of diabetes and diabetic neuropathy.

This should go under the "does not have a clue" heading. Incredible as it seems, researchers at Washington University School of Medicine in Missouri are currently evaluating BOTOX® injections to help treat foot ulcers.[65] Botox

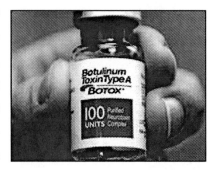

injections are a diluted form of botulism that will paralyze the specified muscle area.

Botulinum toxin is made by the bacteria *Clostridium botulinum*. The bacteria themselves (and their spores) are harmless, but the toxin is considered one of the most lethal known poisons, one that has been a principle agent in biological warfare.[66] It binds to nerve endings where they join muscles, leading to weakness or paralysis. Recovery from botulism occurs when the nerves grow new endings, which can take months, according to the FDA.[67] Choosing highly toxic options to treat illness has no medical merit when there are infinitely safer treatments like magnesium chloride that is so safe that it helps prevent the development of foot ulcers and diabetic neuropathy in the first place.

> *Rapid increase of magnesium stores are necessary*
> *in some cases and may be lifesaving for diabetics*
> *as they are for other patients in emergency rooms.*

The nutritional effects of restoring magnesium to normal levels, may be profoundly preventative, going a long way to protecting the children of the future from early onset of both diabetes and the complications that come from it. The safety profile of magnesium chloride as a restoration modality is extraordinary compared to today's pharmaceutical drugs. It is only with severe renal insufficiency that problems have been observed with magnesium treatments. The elderly are at risk of magnesium toxicity only because of possible decreased renal function so caution is necessary.

Special Note: While Dr. Mansmann makes a strong case for high doses of magnesium, it cannot be ignored that GLA has also been recognized for its ability to stop and/or reverse peripheral neuropathy and is endorsed by Dr. Robert C. Atkins, of the famous Atkins diet, which many diabetics follow.

Dr. Atkins says, "Science has established rather conclusively that GLA halts the otherwise inevitable advance of nerve damage caused by diabetes. GLA helps the nerves to heal. As one study of 111 patients showed, people with either form of diabetes, Type I or Type II, can benefit, using a dose as small as 480 mg of GLA per day.[68] Other research suggests that the fatty acid may even prevent the nerve deterioration from starting up.[69]

Some kind of abnormality in fatty acid metabolism is very likely involved in the development of diabetic complications and maybe even the development of diabetes itself. People who have the disease seem unable to make GLA from dietary fats and therefore may suffer from an insufficiency of PGE_1, (Prostaglandin E_1, a beneficial hormone-like compound). Coincidentally enough, this substance can potentiate the work of insulin and exerts insulin like actions of its own. Therefore diabetics need all the PGE_1 that GLA can help them make."

References

1   v57, *Better Nutrition for Today's Living*, March '95, p34. http://www.mgwater. com/articles.shtml

2   Department of Internal Medicine, Overlook Hospital, Summit, NJ, USA. Hypomagnesemia has long been known to be associated with diabetes mellitus. Mather et al confirmed the presence of hypomagnesemia in nearly 25% of their diabetic out-patients. Low serum magnesium level has been reported in children with insulin-dependent diabetes and through the entire spectrum of adult type I and type II diabetics regardless of the type of therapy. Hypomagnesemia has been correlated with both poor diabetic control and insulin resistance in nondiabetic elderly patients. Hypomagnesemia and diabetes mellitus. A review of clinical implications. Tosiello L; *Arch Intern Med.* 1996 Jun 10;156(11):1143-8. (www.ncbi.nlm.nih.gov/)

3   Barbagallo, Mario et al. Effects of vitamin E and Glutathione on Glucose Metabolism: Role of Magnesium; (*Hypertension.* 1999;34:1002-1006.) American Heart Association (www.ncbi.nlm.nih.gov/)

4   Hua H. et al: Magnesium transport induced ex vivo by a pharmacological dose of insulin is impaired in non-insulin-dependent diabetes mellitus.Magnes Res. 1995, Dec; *Magnes Res*. 1995 Dec;8(4):359-66. PMID: 8861135 [PubMed - indexed for MEDLINE]

5   A tendency for magnesium deficiency in patients with diabetes mellitus is well-established. Glucosuria-related hypermagnesiuria, nutritional factors and hyperinsulinaemia-related hypermagnesiuria all can contribute. The plasma magnesium level has been shown to be inversely related to insulin sensitivity. Magnesium supplementation improves insulin sensitivity as well as insulin secretion in patients with type 2 diabetes. Nevertheless, no beneficial effects of oral magnesium supplementation has been demonstrated on glycaemic control either in patients with diabetes type 1 or 2. Oral magnesium supplementation reduced the development of type 2 diabetes in predisposed rats. There are some indications that magnesium decreases blood pressure, but negative results have been observed in trials that were, however, not designed to test effect on blood pressure as primary parameter. Patients with (severe) retinopathy have a lower plasma magnesium level compared to patients without retinopathy and a prospective study has shown the plasma magnesium level to be inversely related to occurrence or progression of retinopathy. Further study on magnesium (supplementation) is warranted in the prevention of type 2 and of (progression of) retinopathy as well as a means to reduce high blood pressure. Magnesium in Diabetes Mellitus; de Valk HW. *Neth J Med.* 1999 Apr;54(4):139-46.

6   Lower serum magnesium levels are associated with more rapid decline of renal function in patients with diabetes mellitus type 2.Clin Nephrol. 2005 Jun;63(6):429-36. PMID: 15960144

7   Diabetic ketoacidosis (DKA) is a dangerous condition that can cause you to lose consciousness. If untreated, it can be fatal. This is a diabetic crisis state, and can quickly lead to fatality, including cerebral edema, most often seen in children. It is also common in DKA to have severe dehydration and significant alterations of the body's blood chemistry. Diabetic ketoacidosis is a triad of hyperglycemia, ketonemia and acidemia. (ketones and acid in the bloodstream) Major components of the pathogenesis of diabetic ketoacidosis are reductions in effective concentrations of circulating insulin

and concomitant elevations of counterregulatory hormones (catecholamines, glucagon, growth hormone and cortisol). These hormonal alterations bring about three major metabolic events: (1) hyperglycemia resulting from accelerated gluconeogenesis and decreased glucose utilization, (2) increased proteolysis and decreased protein synthesis and (3) increased lipolysis and ketone production.

8   Diabetic ketoacidosis: Check your ketones; From MayoClinic.com Special to CNN.com (www.cnn.com/HEALTH/library/DA/00064.html)

9   This article exemplifies the AAFP 1999 Annual Clinical Focus on management and prevention of the complications of diabetes. Diabetic ketoacidosis is an emergency medical condition that can be life-threatening if not treated properly. The incidence of this condition may be increasing, and a 1 to 2% mortality rate has stubbornly persisted since the 1970s. Diabetic ketoacidosis occurs most often in patients with Type I diabetes (formerly called insulin-dependent diabetes mellitus); however, its occurrence in patients with Type II diabetes (formerly called non¬insulin-dependent diabetes mellitus), particularly obese black patients, is not as rare as was once thought. The management of patients with diabetic ketoacidosis includes obtaining a thorough but rapid history and performing a physical examination in an attempt to identify possible precipitating factors. The major treatment of this condition is initial rehydration (using isotonic saline) with subsequent potassium replacement and low-dose insulin therapy. The use of bicarbonate is not recommended in most patients. Cerebral edema, one of the most dire complications of diabetic ketoacidosis, occurs more commonly in children and adolescents than in adults. Continuous follow-up of patients using treatment algorithms and flow sheets can help to minimize adverse outcomes. Preventive measures include patient education and instructions for the patient to contact the physician early during an illness. (Am Fam Physician 1999;60:455-64.)ABBAS E. KITABCHI, PH.D., M.D., and BARRY M. WALL, M.D. University of Tennessee, Memphis, College of Medicine Memphis, Tennessee (www.aafp.org/afp/990800ap/455. html)

10   Chronic Renal Failure (Chronic Renal Insufficiency, Kidney Failure, Renal Insufficiency) (CRF) Irreversible, progressive impaired kidney function. The early stage, when the kidneys no longer function properly but do not yet require dialysis, is known as Chronic Renal Insufficiency (CRI). CRI can be difficult to diagnose, as symptoms are not usually apparent until kidney disease has progressed significantly. Common symptoms include a frequent need to urinate and swelling, as well as possible anemia, fatigue, weakness, headaches and loss of appetite. As the disease progresses, other symptoms such as nausea, vomiting, bad breath and itchy skin may develop as toxic metabolites, normally filtered out of the blood by the kidneys, build up to harmful levels. Over time (up to 10 or 20 years), CRF generally progresses from CRI to End-Stage Renal Disease (ESRD, also known as Kidney Failure). Patients with ESRD no longer have kidney function adequate to sustain life and require dialysis or kidney transplantation. Without proper treatment, ESRD is fatal.

11   Source: *News Day* (www.newsday.com/news/health/ny-usdiab074502530nov07,0,585 4471.story?coll=ny-health-headlines

12   Hiller TA, Pedula KL. Complications in young adults with early-onset type 2 diabetes: losing the relative protection of youth. Center for Health Research, Kaiser Permanente Northwest/Hawaii, Portland, Oregon. Diabetes Care. 2003 Nov;26(11):2999-3005. CONCLUSIONS: Early-onset Type II diabetes appears to be a

more aggressive disease from a cardiovascular standpoint. Although the absolute rate of cardiovascular disease (CVD) is higher in older adults, young adults with early-onset type 2 diabetes have a much higher risk of CVD relative to age-matched control subjects. Source: paktribune.com/news/index.php?id=124683

13   American Diabetes Association (www.diabetes.org)

14   *Diabetes Care* 28:1175–1181, 2005.

15   New York Times. January 9, 2006

16   Derrick Z. Jackson, Diabetes and the trash food industry. Boston Globe. January 11, 2006

17   New York Times. January 11, 2006.

18   Increasing rates of Type II diabetes worldwide suggest that diabetes may be caused by environmental toxins. Cadmium is a widespread environmental pollutant that accumulates in the pancreas and exerts diabetogenic effects in animals. To test the hypothesis that exposure to cadmium is associated with impaired fasting glucose and Type II diabetes, we examined the associations between urinary cadmium and the prevalence of impaired fasting glucose (prediabetes) and diabetes in the Third National Health and Nutrition Examination Survey (NHANES III). In this large cross-sectional study, urinary cadmium levels are significantly and dose-dependently associated with both impaired fasting glucose and diabetes. These findings, which require confirmation in prospective studies, suggest that cadmium may cause prediabetes and diabetes in humans. Urinary Cadmium, Impaired Fasting Glucose, and Diabetes in the NHANES III Gary G. Schwartz, PHD et.al; *Diabetes Care* 26:468-470, 2003   (http://care.diabetesjournals.org/cgi/content/full/26/2/468)

19   Metals such as iron, mercury, arsenic, lead and possibly aluminum may play a role in the actual destruction of beta cells through stimulating an auto-immune reaction to them after they have bonded to these cells in the pancreas. What we will focus on here though is the fact that insulin has three sulfur-containing cross-linkages and the insulin receptor has a tyrosine kinase-containing sulfur bond, which are the preferred targets for binding by both mercury and lead. Should mercury attach to one of these three sulfur bonds it will interfere with the normal biological function of the insulin molecule. In reality there is no "should" about it, the average adult inhales many trillions of mercury atoms a day from a mouth full of amalgam, fish provide trillions more, the air more, and in children, vaccines provide one day surges of vast trillions of mercury molecules in the form of ethyl-mercury, which is vastly more toxic than metallic mercury. Insulin molecules are directly assaulted, as are insulin receptor sites. *The Hun Hordes of Mercury*; (http://imva.info/)

20   Toxicity of Fluoride to Diabetic Rats. C.A.Y. Banu Priya et al; International Society for Fluoride Research; FLUORIDE 30 (1)1997, pp 51 - 58 (www.fluoride-journal.com/97-30-1/301-51.htm)

21   Vandiver J, "Chicken Feed," *Daily Times* (Salisbury, Md.), January 4, 2004.

22   Tseng CH, Tseng CP, Chiou HY, Hsueh YM, Chong CK, Chen CJ. Epidemiologic evidence of diabetogenic effect of arsenic. *Toxicol Lett*. 2002 Jul 7;133(1):69-76.

23   Mahfuzar Rahman et al. Division of Occupational and Environmental Medicine, Department of Health and Environment, Faculty of Health Science Linkoping University Sweden. Department of Occupational and Environmental Health(DOEH),

National Institute of Preventive and Social Medicine (NIPSOM), Mohakhali, Dhaka-1212 Bangladesh. *American Journal of Epidemiology* 1998; Vol. 148, No.2: 198-203 The crude prevalence ratio for diabetes mellitus among keratotic subjects exposed to arsenic was 4.4 (95% confidence interval 2.5-7.7) and increased to 5.2 (95% confidence interval 2.5-10.5) after adjustment for age, sex, and body mass index.

24   Source: www.ehponline.org/docs/2003/6407/abstract.html.

25   A solution of alloxan at 2% diluted in saline at 0.9% was administered to the animals in a single dose corresponding to 40 mg of alloxan per kg of animal weight injected into their penial vein. Alloxan induces irreversible diabetes mellitus after 24 hours following its administration and the condition proves to be chronic by laboratory tests after seven days. Experimental Model of Induction of Diabetes Mellitus in Rats; Acta Cir. Bras. vol.18 no.spe S o Paulo 2003 (www.scielo.br/ scielo.php?pid=S0102-86502003001100009&script=sci_arttext&tlng=en)

26   Researchers who are studying diabetes commonly use the chemical to induce the disorder in lab animals. Unfortunately, most consumers are unaware of alloxan and its potentially fatal link to diabetes because these facts are not well publicized, are hidden by FDA approval, and certainly doctors and the food industry are not informing parents that they and their children are being poisoned by white flour containing alloxan. Diabetes and Chemical Poisoning. Source: (http://imva.info/)

27   *Consumer Reports* (Feb. 2006):

28   Genetically Engineered Food Biotech, Biotechnology, GMO, Genetically Modified (www.organicconsumers.org/gelink.html)

29   Health Hazards of Genetically Manipulated Foods; www.soyinfo.com/haz/gehaz.shtml

30   Dr. Irina Ermakova added flour from a GM soya bean - produced by Monsanto to be resistant to its pesticide, Roundup - to the food of female rats, starting two weeks before they conceived, continuing through pregnancy, birth and nursing. Others were given non-GM soya and a third group was given no soya at all. She found that 36% of the young of the rats fed the modified soya were severely underweight, compared to 6 per cent of the offspring of the other groups. More alarmingly, a staggering 55.6% of those born to mothers on the GM diet perished within three weeks of birth, compared to 9 per cent of the offspring of those fed normal soya, and 6.8 per cent of the young of those given no soya at all. (www.organicconsumers.org/ge/babies010906.cfm)

31   Malatesta M, Caporaloni C, Rossi L, Battistelli S, Rocchi MBL, Tonucci F, Gazzanelli G (2002) Ultrastructural analysis of pancreatic acinar cells from mice fed on genetically modified soybean. *Journal of Anatomy* 201:409-415

32   Mary L. Johnson, RN, CDE, International Diabetes Center, Minneapolis, Minnesota, and colleagues[12] evaluated 206 persons for the microvascular complications of peripheral neuropathy, retinopathy, and nephropathy. On average, subjects had diabetes for approximately 10 years and demonstrated good glucose control (hemoglobin A1C 7.3% ±1.4%). 48% of study subjects had microvascular complications, even though they had generally good glycemic control and a relatively short duration of diabetes. Diabetic nephropathy was identified in 20% of the study population; retinopathy was identified in 11%; and symptoms of diabetic peripheral neuropathy were identified in 63%. Diagnosis with a 10-g Semmes-Weinstein monofilament only identified 16% of patients with neuropathy. However, over 30% of

the subjects exhibited sensory deficits after clinical examination. www.medscape.com/viewarticle/508218

33   Additional symptoms that may be associated with this disease: swallowing difficulty, speech impairment, loss of function or feeling in the muscles, muscle contractions, muscle atrophy, uncoordinated movement, dysfunctional movement, joint pain, hoarseness or changing voice, fatigue, facial paralysis, eyelid drooping, bowel or bladder dysfunction, breathing difficulty. From the National Institutes of Health (NIH) National Library of Medicine (NLM) MEDLINEplus Medical Encyclopedia.

34   WHO, 2002; (www.who.int/mediacentre/factsheets/fs138/en/print.html)

35   A 10-year clinical study that involved 1,441 volunteers with insulin-dependent diabetes (IDDM) was recently completed by the National Institute of Diabetes and Digestive and Kidney Diseases. The study proved that keeping blood sugar levels as close to the normal range as possible slows the onset and progression of nerve disease caused by diabetes. The Diabetes Control and Complications Trial (DCCT) studied two groups of volunteers: those who followed a standard diabetes management routine and those who intensively managed their diabetes. Persons in the intensive management group took multiple injections of insulin daily or used an insulin pump and monitored their blood glucose at least four times a day to try to lower their blood glucose levels to the normal range. After 5 years, tests of neurological function showed that the risk of nerve damage was reduced by 60% in the intensively managed group. People in the standard treatment group, whose average blood glucose levels were higher, had higher rates of neuropathy. Although the DCCT included only patients with IDDM, researchers believe that people with noninsulin-dependent diabetes would also benefit from maintaining lower levels of blood glucose.

36   2005 Survey Results Fact Sheet; (www.diabetesincontrol.com/modules.php?name=News&file=article&sid=3253)

37   Episodes of hypoglycemia will also cause the release of counterregulatory hormones which act to elevate the blood sugars for the following 24-48 hours, further complicating the issue of "good control". Even the ADA states recognition of the fact that it is difficult to maintain blood glucose levels within their recommendations, for any appreciable length of time. Yet target levels have been lowered over the past few years to levels that may be causing more hypoglycemia than ever before. Studies in children show a high rate of hypoglycemia at night for sustained periods of time. Unnoticed by caregivers. Hypoglycemia can occur before the diagnosis of diabetes, and during treatment with oral antihyperglycemic drugs or insulin injections.

38   Insulin induced neuropathy has been reported previously in people with diabetes treated with insulin, and subsequently reported in patients with insulinomas. However, neuropathy caused by rapid glycaemic control in patients with poorly controlled diabetes with chronic hyperglycaemia is not a widely recognised entity among clinicians worldwide. It is expected that this phenomenon of paradoxical complication of neuropathy in the face of drastic decreases in glycosylated haemoglobin concentrations will assume greater importance with clinicians achieving glycaemic targets at a faster pace than before. Under-recognised paradox of neuropathy from rapid glycaemic control. Postgrad Med J. 2005 Feb;81(952):103-7.Leow MK, Wyckoff J. Department of Endocrinology, Division of Medicine, Tan Tock Seng Hospital, Singapore. mleowsj@massmed.org (www.ncbi.nlm.nih.gov/entrez/query.fcgi?cmd=Retrieve&db=pubmed&dopt=Abstract&list_uids=15701742&itool=iconabstr&query_

hl=6&itool=pubmed_docsum)

39   The effect of sustained insulin-induced hypoglycemia on peripheral nerve function and structure was examined in rats. After a period of hypoglycemia (less than 2.5 mmol/L) of at least 72 h, axonal degeneration and reduction of the maximal amplitude of the evoked muscle action potential occurred, the two abnormalities being correlated negatively (r = -0.99, 2P = 0.00097). One of five rats developed paresis of both hindlegs as well as nerve damage and perikaryal alterations of lower motor neurons. Peripheral neuropathy in rats induced by insulin treatment: P Sidenius and J Jakobsen ; Diabetes, Vol 32, Issue 4 383-386, Copyright 1983 by American Diabetes Asssociation. (http://diabetes.diabetesjournals.org/cgi/content/abstract/32/4/383)

40   Source: http://diabetes.niddk.nih.gov/dm/pubs/statistics/index.htm#11

41   Source: Joslin EP. The menace of diabetic gangrene. N Engl J Med 1934; 211: 16–20

42   Although the treatment of diabetes has become increasingly sophisticated, with over a dozen pharmacological agents available to lower blood glucose, a multitude of ancillary supplies and equipment available, and a clear recognition by health care professionals and patients that diabetes is a serious disease, the normalization of blood glucose for any appreciable period of time is seldom achieved . In addition, in well-controlled so-called "intensively" treated patients, serious complications still occur, and the economic and personal burden of diabetes remains. Furthermore, microvascular disease is already present in many individuals with undiagnosed or newly diagnosed type 2 diabetes . (http://care.diabetesjournals.org/cgi/content/full/diacare;27/suppl_1/s47)

43   Andrew J.M. Boulton. Professor of Medicine, University of Manchester, and Department of Medicine, Manchester Royal Infirmary, Manchester, UK See: www.d4pro.com/idm/site/leading_article5.htm

44   The Ten Commandments of the Diabetic Foot BMJ: 2005;331:1497 (24 December) doi:10.1136/bmj.331.7431.1497

45   Source: diabetes.niddk.nih.gov/dm/pubs/alternativetherapies/

46   Total serum magnesium was reduced in the high-fructose group compared with control or high-fructose plus magnesium-supplemented groups. Blood pressure and fasting insulin levels were also lower in the magnesium-supplemented group. These results suggest that magnesium deficiency and not fructose ingestion per se leads to insulin insensitivity in skeletal muscle and changes in blood pressure. Dietary magnesium prevents fructose-induced insulin insensitivity in rats. Batan et.al; *Hypertension*. 1994 Jun;23(6 Pt 2):1036-9.

47   Endocrinology & Metabolism Clinics of North America. 24(3):623-41, 1995 Sep.

48   Carper, J. "Mighty Magnesium: This overlooked nutrient fights against heart disease, pain and diabetes." *USA Weekend*. 2002 Aug 30-Sept 1.

49   Magnesium Deficiency Linked to Type II Diabetes (www.newstarget.com/006121.html) Studies conducted at Harvard University indicate that people who have high levels of magnesium in their blood are less likely to develop Type II diabetes or insulin resistance than those with lower levels. Studies in Mexico have also found an alleviation of diabetes symptoms in patients who took dietary supplements containing magnesium. Original Source: www.health24.com/dietnfood/General/15-

742-775,31268.asp

50   "Diabetologia" 36(8):767-70, 1993

51   Low serum magnesium levels and foot ulcers in subjects with Type II diabetes. Rodriguez-Moran M, Guerrero-Romero F. Arch Med Res. 2001 Jul-Aug;32(4):300-3.

52   S. E. BROWNE. The Case for Intravenous Magnesium Treatment of Arterial Disease in General Practice. *Journal of Nutritional Medicine* (1994) 4, 169-177

53   Herbert C. Mansmann Jr. MD. Honorary Professor of Pediatrics. P.O. Box 791, Rangeley, ME 04970 Associate Professor of Medicine (1968-03) Director of the Magnesium Research. Laboratory (1989-03) Thomas Jefferson University (www. magnesiumresearchlab.com)

54   Source: http://health.groups.yahoo.com/group/MagnesiumResearchLab/message/2863

55   Source: http://magnesiumresearchlab.com/Diabetes-and-Mg-5-11-04.htm

56   The effect of magnesium supplementation in increasing doses on the control of Type II diabetes. *Diabetes Care*. 1998 May;21(5):682-6.

57   Source: http://magnesiumresearchlab.com/Diabetes-and-Mg-5-11-04.htm

58   See www.MagnesiumForLife.com for full information on transdermal magnesium chloride mineral therapy. And go to http://www.globallight.net to see the recommended natural seawater product with the highest concentration and lowest toxicity.

59   Long term magnesium supplementation influences favorably the natural evolution of neuropathy in magnesium-depleted Type I diabetic patients (T1dm); De Leeuw et al; *Magnes Res.* 2004 Jun; 17(2):109-14

60   Source: mgwater.com/

61   Because fluoride is excreted through the kidney, people with renal insufficiency would have impaired renal clearance of fluoride (Juncos and Donadio 1972) Elderly people are more susceptible to fluoride toxicity.

62   Statins and peripheral neuropathy; U. Jeppesen , D. Gaist , T. Smith S. H. Sindrup *European Journal of Clinical Pharmacology* Volume 54, Number 11;835 - 838 January 1999

63   The Peripheral Neuropathy Caused by Statins Petition to Pharmaceutical Researchers and Manufacturers of America and companies listed was created by DrugIntel Statin Users with Neuropathy and written by John Lehmann. "We users of statin drugs have experienced some of the symptoms listed below [1] that characterize peripheral neuropathy (damage to nerves outside the brain). Medical research published in peer-reviewed journals has shown that statins are able to cause peripheral neuropathy or a syndrome that is very similar to it. We petition the pharmaceutical manufacturers of statins [2] to: 1. Notify patients (past, current, and prospective users of statins) and health care professionals (physicians, pharmacists, nurses, physicians' assistants) of the risk associated with statin use and what to do once the first signs and symptoms of neuropathy have appeared. 2. Sponsor and perform research on how statins cause neuropathy. 3. Sponsor and perform clinical research on how to cure and reverse the neuropathy caused by statins. 4. Perform clinical research and recommend the best drug treatments to mitigate the pain and make other symptoms of statin-induced neuropathy more tolerable. 5. Proactively offer reparation to statin users who have suffered

neuropathy. The petition will be presented to the Pharmaceutical Researchers and Manufactuers Association and to the Medical Affairs Departments of the companies listed, as well as any additional companies that may be identified as relevant over time http://www.petitiononline.com/Statins/petition.html

64   Statins and risk of polyneuropathy

D Gaist, MD PhD, U Jeppesen, M Andersen, *LAG Neurology* 2002;58:1333-1337 © 2002 American Academy of Neurology Statins and risk of polyneuropathy.

65   Participants receive injections of the toxin in six places in the calf muscle and then the leg is put into a cast. The idea is that this will help prevent pressure on the ball of the foot during walking. The ball if the foot is the area most affected by foot ulcers and allowing an ulcer to heal completely helps prevent recurrence. (www.diabetes-and-diabetics.com/about-diabetes/diabetic-complications-02.php)

66   Botulinum toxin has been a concern as a potential biological warfare agent since World War II. In response to concerns about Germany's botulinum toxin research, the United States and Great Britain developed countermeasures against the toxin before the invasion of Europe. More recently, Iraq has been accused of producing large amounts of botulinum toxin for use as a biological warfare agent. The extreme toxicity of botulinum toxins and the ease of production, transport, and delivery make this an agent of extreme bioterrorism concern.

67   Overview of Botulism: www.cidrap.umn.edu/cidrap/content/bt/botulism/biofacts/botulismfactsheet.html

68   Keen, H., et al., *Diabetes Care*, 1993; 16: 8-15.

69   Jamal, G., *Diabetic Medicine*, 1994; 11(2): 145-49.

11

||||||||||||||||||||||||||||||||||||||||||||||||||||||||||||||||||||||

Magnesium and Strokes

Stroke is the third leading cause of death in the United States and the most common cause of adult disability. An ischemic stroke occurs when a cerebral vessel occludes, obstructing blood flow to a portion of the brain. Each year, 700,000 Americans suffer a stroke. If they do not die on the spot, nearly 25% of them will die in a year from lack of appropriate treatment. Those 1.2 million Americans who have survived strokes now report serious disabilities that affect daily living. How we treat strokes is very important because the list of disabilities of patients age 65 or over, six months after they had suffered their stroke, is appalling.[1]

> 50% suffer paralysis on one side of their body.
>
> 35% have symptoms of depression
>
> 30% can't walk without assistance
>
> 26% need help with daily activities
>
> 26% are living in nursing homes
>
> 19% have speech or language problems

The cost to care for stroke victims in America, this year, alone, is $54 billion. Most think that taking aspirin every day will prevent platelet aggregation (clot formation) and help prevent stroke. However, the truth is that aspirin may prevent stroke in only 3 of 100 women and does not seem to prevent stroke in men at all.[2] Aspirin is not the correct preventive treatment for stroke and that point is driven home when we consider that aspirin causes gastrointestinal bleeding in 8 out of 1,000 people and is sometimes fatal.

Ischemia leads to excessive activation of excitatory amino acid receptors, accumulation of intracellular calcium, and release of other toxic products that cause cellular injury.

Dr. James Howenstine writes, "The use of aspirin has become widely accepted in the United States as an important measure to prevent heart attacks and strokes. Estimates suggest that 20,000,000 persons are taking aspirin daily for prevention of vascular accidents. The evidence upon which this decision to recommend aspirin was made is not very solid. Millions of patients with heart attacks, strokes, angina pectoris, diabetes and risk factors for vascular disease have been encouraged by their physicians to take aspirin to prevent heart attacks and strokes. Four early studies using aspirin to prevent heart attacks had shown no benefit.[3,4]

"Then along came a study on U.S. physicians, which used Bufferin (which is comprised of aspirin and magnesium). This study showed no reduction in fatal heart attacks and no improvement in survival rate, but there was a 40 % decrease in the number of non fatal heart attacks. The magnesium was ignored and there was a prompt extensive institution of aspirin for prevention of heart attacks. The benefits of magnesium in treating heart disease include the well known decrease in ischemic heart disease and sudden death found in communities drinking hard water (magnesium containing), powerful prevention of platelet clumping (clot prevention) known to be caused by magnesium, strong blood vessel dilating properties of magnesium, and effective action to block dangerous heart rhythms in persons taking magnesium. The decrease in number of heart attacks probably resulted from the magnesium in Bufferin."[5]

Magnesium deficiency can cause metabolic changes that may contribute to heart attacks and strokes.[6]
National Institutes of Health

Once you get a stroke you have limited treatment options, one of which is taking more aspirin. If the dose is too high though you run the risk of GI bleeding, ringing in the ears, and hearing loss. There are other medications called anti-platelet aggregators that may reduce risk for further stroke, death, or heart attack by 9%, but the side effects escalate to diarrhea, skin rash, and agranulocytosis (a condition that wipes out your white blood cells). Most of the non-aspirin anti-platelet drugs cost three to four dollars per day or $1,000 to $1,500 a year for the rest of your life.

High blood pressure (hypertension) is one of the major vascular disorders that magnesium can help.[7]
Dr. Jay Cohen

Statin drugs are also prescribed to stroke patients for a potential 25-30% reduction of future stroke. The cost of these pills also ranges from three to four dollars a day. Another group of drugs given to stroke patients is anti-hypertensives. Often the stress of a stroke will elevate blood pressure, or it may even be one of the primary causes. Taking these drugs along with an anti-platelet drug may reduce the relative risk of stroke, death, and myocardial infarction by 32%. The price for many of the anti-hypertensives is also about three dollars a day or $1000 per year.

Another class of treatments are blood thinners called thrombin inhibitors. The main one is Warfarin, which is the main ingredient in rat poison. It acts by causing fatal hemorrhage in rats. In humans the dosage has to be adjusted very carefully by taking regular blood tests. This drug is often necessary immediately after a stroke to prevent clot formation in the heart and to diminish the incidence of heart arrhythmias.

Bleeding and bruising are the main side effects of blood thinners. This is a non-patented drug and a year's supply may only be about $200. However, the weekly doctor's appointments and blood tests make blood thinning a highly lucrative therapy for modern medicine.

The most commonly studied neuroprotective agents for acute stroke block the N-methyl-D-aspartate (NMDA) receptor. Dextrorphan, a noncompetitive NMDA antagonist and metabolite of cough suppressant, was the first NMDA antagonist studied in human stroke patients. Unfortunately, dextrorphan caused hallucinations and agitation; it also produced hypotension, which limited use. Magnesium is an agent with actions on the NMDA receptor and a low incidence of side effects. It may reduce ischemic injury by increasing regional blood flow, antagonizing voltage-sensitive calcium channels, and blocking the NMDA receptor.

Using various mechanisms, neuroprotective agents attempt to save ischemic neurons in the brain from irreversible injury. Studies in animals indicate a period of at least four hours after onset of complete ischemia in which many potentially viable neurons exist in the ischemic penumbra.[8] With magnesium treatments, the trend toward a better functional outcome at 30 days in patients is seen when treatments are started within twenty-four hours from onset versus controls.[9]

Intravenous magnesium sulfate administration during the hyper acute phase of stroke was shown to be safe in a small, open-label pilot trial, in which more than 70% of patients were treated less than 2 hours from symptoms onset. Dramatic early recovery was achieved in 42% of patients, and good functional outcome (modified Rankin scale </= 2) at 90 days post treatment was achieved by 69% of all patients and in 75% treated within 2 hours.[10]

Dr. Carolyn Dean says, "Magnesium is important in lowering blood pressure, keeping the heart muscle from going into spasm, and lowering cholesterol (by the same mechanism as statin drugs) but it can help heal the damage in the brain caused by a stroke." Magnesium, an important cofactor in metabolism and protein synthesis, joins into a complex with adenosine triphosphate. Magnesium acts as a noncompetitive NMDA receptor blocker; it inhibits the release of excitatory neurotransmitters at the presynaptic level and blocks voltage-gated calcium channels. Moreover, it has been shown to suppress anoxic depolarization and cortical spreading depression — both potential targets for neuroprotective treatment. Magnesium also exerts vascular effects, such as boosting vasodilatation, increasing the cardiac output, and prolonging bleeding time.

*In my practice the use of magnesium in the
early stages of a stroke has rendered the best results
for my patients who have the greatest deficits.*

Dr. Al Pinto

Intravenous magnesium sulphate protects ischemic neurons *in vitro* and *in vivo* in standard animal experimental stroke models, including global 4-vessel forebrain ischaemia (Tsuda et al 1991), permanent middle cerebral artery occlusion (Izumi 1991) and direct NMDA injection (McDonald et al 1990).

Neuroprotection may be due to a number of properties of magnesium: vasodilatation by magnesium sulphate increases blood flow to the ischemic cortex (Chi et al 1990) whilst increasing cardiac output; it prevents cerebral vasospasm (Kemp et al 1993); it is the endogenous non-competitive blocker of NMDA receptors (Nowak et al 1984, Harrison and Symmonds 1985), a property which may be responsible for its efficacy as an anti-convulsant (The Eclampsia Trial Collaborative Group 1995); and it antagonizes calcium entry to cells via multiple channels (Iseri and French 1984).[11]

Magnesium chloride has several advantages over other neuroprotective agents currently available or in development. A gallon and a half of magnesium chloride a year, at a maximum cost of $180, would do more to prevent strokes safely, (heart attacks, and probably cancer if combined with selenium) without side effects, than any other single medicine.

A gallon, used transdermally after a stroke, over a two or three month period, will reduce disability; and when combined with acupuncture and other nutritional and herbal interventions will, in due time, prove to be the treatment of choice.

References

1   Source: www.msnbc.msn.com/id/4409271/site/newsweek/

2   Source: www.emedicine.com/pmr/topic187.htm

3   *British Medical Journal* 1974 vol. 1 p. 436, *Lancet* Vol. 2 p. 1313, R. Levy *JAMA* Feb 15, 1980

4   R. Peto *British Medical Journal* 1988 vol. 296 pg. 313-6

5   Source: www.newswithviews.com/Howenstine/james10.htm

6   Source: http://healthlink.mcw.edu/article/965927519.html

7   *The Magnesium Solution for High Blood Pressure* by Dr. Jay Cohen

8   One action of neuroprotective agents limits acute injury to neurons in the penumbra region or rim of the infarct after ischemia. Neurons in the penumbra are less likely to suffer irreversible injury at early time points than are neurons in the infarct core. Many of these agents modulate neuronal receptors to reduce release of excitatory neurotransmitters, which contribute to early neuronal injury.

9   Lampl Y, Gilad R, Geva D, Eshel Y, Sadeh M. Intravenous administration of magnesium sulfate in acute stroke: a randomised double-blind study. *Clin Neuropharmacol*. 2001;24:11-15. Abstract

10   Saver JL, Kidwell C, Eckstein M, et al. Prehospital neuroprotective therapy for acute stroke: results of the field administration of stroke therapy -- magnesium (FAST-MAG) pilot trial. *Stroke*. 2004;35:106-108.

11   Source: www.neurosurgery-neff.com/IMAGES.html

12

||||||||||||||||||||||||||||||||||||||||||||||||||||||||||||||||||||||||

Magnesium, Violence and Depression

In February of 2006 the New York Times reported that, "While violent crime has been at historic lows nationwide and in cities like New York, Miami and Los Angeles, it is rising sharply in many other places across the country. And while such crime in the recent past was characterized by battles over gangs and drug turf, the police say the current rise in homicides has been set off by something more bewildering: petty disputes that hardly seem the stuff of fist fights, much less gunfire or stabbings. Suspects tell the police they killed someone who "disrespected" them or a family member, or someone who was 'mean mugging' them, which the police loosely translate as giving a dirty look."[1]

"Police Chief Nannette H. Hegerty of Milwaukee calls it 'the rage thing,'" the Times reported, "We're seeing a very angry population, and they don't go to fists anymore, they go right to guns," she said. "When we ask, 'Why did you shoot this guy?' it's (because), 'He bumped into me,' 'He looked at my girl the wrong way,'" said

Police Commissioner Sylvester M. Johnson of Philadelphia. "It's not like they're riding around doing drive-by shootings. It's arguments — stupid arguments over stupid things."

While arguments have always made up a large number of homicides, the police say the trigger point now comes faster. In robberies, Milwaukee's Chief Hegerty said, "Even after the person gives up, the guy with the gun shoots him anyway. We didn't have as much of that before."

What could be driving such a surge in violence? We can easily suspect that with a reported magnesium deficiency of almost 70% of the American population, we are witnessing the extreme end of the most severe deficiencies in the population.

The two most basic requirements for the normal operation of our brain are a sufficient energy supply and an optimal presence of biochemicals involved in transmitting messages. Magnesium is crucial in both the production of energy and neurotransmitters, not to mention the integrity of the blood brain barrier. It is bedrock science that connects magnesium to neurological disorders.[2]

> *Magnesium deficiency causes serotonin-deficiency*
> *with possible resultant aberrant behaviors,*
> *including depression, suicide or irrational violence.*
> Paul Mason

Magnesium of course is not the only nutrient whose deficiency is leading to broad problems of mind and emotion. Zinc is also an important mineral and is involved with psychiatric disorders. Over 90 metallo-enzymes require zinc and the functioning of the brain is dependent on adequate levels of zinc. Deficiency can cause amnesia, apathy, depression, irritability, lethargy, mental retardation and paranoia. As it is for magnesium and zinc it is for a host of basic nutrients though it is the mineral deficiencies that are the most important.

Numerous studies conducted in juvenile correctional institutions have reported that violence and serious antisocial behavior have been cut almost in half after implementing nutrient-dense diets.

But health officials and the pharmaceutical companies want to know nothing about using such simple substances as minerals to help depression or violence. Since the arrival of selective serotonin reuptake inhibitors, antidepressants (SSRIs), and atypical antipsychotics on the market, countless studies have shown the so-called "new generation" of psychiatric drugs to be ineffective and dangerous. Worldwide, sales of anti-psychotics went from $263 million in 1986 to $8.6 billion in 2004 and anti-depressant sales went from $240 million in 1986, to $11.2 billion in 2004. For these two classes of drugs combined, sales went from $500 million in 1986 to nearly $20 billion in 2004, a 40-fold increase, according to Robert Whitaker, best-selling author of *Mad in America*.[3]

Despite a dramatic increase in treatment of psychiatric disorders during the past 10 years, there has been no decrease in the rate of suicidal thoughts and behavior among adults, according to a federal study primarily funded by the National Institute of Mental Health.

The Washington Post

June, 2005

Though it is a complex matrix of causes that cuts across physical, emotional, mental, and spiritual levels of being – it's arguable that a significant portion of the blame for violence and depression can be laid on nutritional causes which are the easiest to correct. It is clear, for example, that magnesium deficiency or imbalance plays a crucial role in the symptoms of mood disorders. Observational and experimental studies have shown an association between magnesium and aggression[4,5,6,7,8], anxiety[9,10,11] ADHD[12,13,14,15], bipolar disorder[16,17], depression[18,19,20,21], and schizophrenia.[22,23,24,25]

Patients who had made suicide attempts
(by using either violent or nonviolent means)
had significantly lower mean CSF
magnesium level irrespective of the diagnosis.[26]

The Department of Family Medicine, Pomeranian Medical Academy, states that dietetic factors can play a significant role in the origin of ADHD and that magnesium deficiency can result in disruptive behaviors.[27] Even a mild deficiency of magnesium can cause sensitiveness to noise, nervousness, irritability, mental depression, confusion, twitching, trembling, apprehension, and insomnia.

Yet as Evelyn Pringle, investigative reporter, lets us know, "Pharma will stop at nothing when it comes to making money off children. On April 25, 2005, the *Ohio Columbus Dispatch* reported an investigation of state Medicaid records that found 18 newborn to 3 years-old babies in Ohio had been prescribed antipsychotic drugs in July 2004. It is a horrible crime and terrible sadness what is being done to the children by pediatricians and psychiatrists who live by the increasingly popular creed to drug the kids with toxic substances."

When the body of a 19-year-old student, Traci Johnson,
was found hanging from a shower rod in the
laboratories of pharmaceuticals giant Eli Lilly,
US officials were quick to announce that the
death could not be linked to a new
anti-depressant drug she was helping to test.[28]

Magnesium ions have nutritional and pharmacologic actions that safely protects against the neurotoxicity of many agents up to and including stress from environmental noise and physical trauma. Magnesium deficiency, even when mild, increases susceptibility to various types of neurological and psychological stressors in both animals and healthy human subjects. "Magnesium deficiency increases susceptibility to the physiologic damage produced by stress. The adrenergic effects of psychological stress induce a shift of magnesium from the intracellular to the extracellular space, increasing urinary excretion and eventually depleting body stores," reports Dr. Leo Galland.[29]

Linus Pauling was one of the first who "highlighted the supremacy of nutrition in correcting abnormalities in the chemical environment of the brain." Nutrients like ascorbic acid, thiamine, niacinamide (vitamin B_3), pyridoxine, vitamin B_{12}, folic acid, magnesium, glutamic acid, and tryptophane were presented by Dr. Pauling as intimately linked to brain function and mental illness.

In addition to strong bodies good nutrition helps us keep our mental health and emotional stability. With the proper diet containing the right nutrients in correct amounts, symptoms of mental illness can be rolled back and treated. Deficiencies in certain necessary nutrients lead to psychotic symptoms and depression while supplementation of other nutrients help attenuate and improve the symptoms of mental illness.

"In 1970 I read about Dr. Abram Hoffer's work and at that time was approached by a friend who had just been stopped from suiciding in a gas oven by her husband. She had her head in there and the gas on. She had also just begun a new drug, for bipolar (manic depression it was named then). She used to be admitted to the local Mental Hospital regularly every year at Spring Time. I began her on hi-dose B_3, magnesium, vitamin C and zinc. Today she is 90 and as bright as a button, very keen mind. In the subsequent 33 years she has only been in mental wards once, and that was when she thought she was cured and didn't have to take her vitamin/mineral formula anymore.

"At that time we were also approached by a young man who had attempted suicide on same drug. The same result we obtained for him. And this was only a small Western cattle and sheep town of 16,000 people. Since, the same results have been obtained in all who have come to me for depression, bipolar, schizophrenia."

-- Michael Sichel, D.O., N.D.
Chittaway Bay, New South Wales, Australia

In 2000, the National Institutes of Health (NIH) listed depression as a sign of magnesium deficiency. NIH defined magnesium deficiency symptoms as having three categories:

> **Early symptoms include** (one or more) irritability, anxiety (including Obsessive Compulsion Disorder and Tourette's syndrome), anorexia, fatigue, insomnia, and

muscle twitching. Other symptoms include apathy, confusion, poor memory, poor attention and the reduced ability to learn.

Moderate deficiency symptoms can consist of the above and possibly rapid heartbeat, irregular heartbeat and other cardiovascular changes (some being lethal).

Severe deficiency symptoms can include one or more of the above symptoms and more severe symptoms including full body tingling, numbness, a sustained contraction of the muscles along with hallucinations and delirium, (including depression) and finally dementia (Alzheimer's disease).

Mild magnesium deficiency appears to be common among patients with disorders considered functional or neurotic and appears to contribute to a symptom complex that includes asthenia, sleep disorders, irritability, hyperarousal, spasm of striated and smooth muscle and hyperventilation.

Normally joy, sadness and grief are parts of everyday life. While a short period of depression in our response to daily problems is normal, a long period of depression and sadness is abnormal. Most depressive episodes are triggered by a stressful personal event such as loss of a loved one or change of circumstances, and depression over a short period is a normal coping mechanism. Long-term stress-induced depression often results when magnesium falls to dangerously low levels in the body. One of the reasons it does this is because the stress itself depletes already meager cellular magnesium stores.

Repletion of deficiency typically reverses any increased stress sensitivity, and pharmacologic loading of magnesium salts orally, parenterally or transdermally induces resistance to neuro-psychologic stressors. If the NIH is aware of this, one wonders why doctors are not currently using magnesium to treat depression and other mental (and physical) disorders asks George Eby, the developer of coldcure.com, who successfully treated himself with magnesium for depression.

George Eby's Testimonial

"I remain truly amazed at the tremendous benefits of magnesium in treating and preventing depression. In particular, I see magnesium as an important research topic for survival considering its limited availability from our Western diets and due to its ability to inexpensively cure and prevent many expensive diseases, life threatening or not.

"I know how bad depression can be, because I spent September of 1999 through April of 2000 in a clinical depression that worsened from the beginning. By Christmas the depression suddenly became much worse, nearly suicidal in intensity, and remained that way for four more months. Never did I think that things could go so wrong with my biochemistry that it would cause me to have suicidal thoughts and tendencies. How wrong I was. I had been taking Zoloft (an antidepressant) since 1987 which seemed to take care of my depression. I lived on Zoloft, but by September of 1999, it stopped working — and I knew that something was really wrong.

"My depression was preceded by many years of major stress from over-work, anxiety, hypomania, fibromyalgia, infrequent panic attacks, anger, stress, poor diet, overwhelming emotional feelings, night time muscle spasms, paranoia, asthma, prickly sensations in hands, arms, chest and lips. I wanted to sleep all day and had trouble getting up in mornings. Occasionally my lips felt that they were going to vibrate or tingle off my face. About 10 years ago, I had a very painful bout with calcium oxalate kidney stones, a recognized sign of magnesium deficiency. A few weeks before I was hospitalized in January of 2000, I had very low energy, mental fogginess, depression with strange suicidal thoughts and I was under enormous stress.

"Now, I can recognize these "mental" symptoms as symptoms of magnesium deficiency and/or calcium toxicity. I was put on nearly every antidepressant drug known and had severe side effects to all of them and felt sicker and sicker. None worked. I lost a lot of weight, and I was extremely constipated. I also had a cardiac arrhythmia.

"On April 12, 2000, I looked like I was dying to several people important in my life. My psychiatrist agreed and took me off all antidepressant medications and put me on a tiny amount of lithium carbonate (150 mg twice a day). Shortly later, I picked up a 1975 copy of *Nutrition Almanac*, McGraw-Hill Book Company, New York, and happened to open it to the magnesium section. I was interested to find that magnesium was low in the serum of people who were suicidal and depressed. The article indicated that magnesium dietary supplements had been effective in treating depression. Also, a person with a magnesium deficiency is apt to be uncooperative, withdrawn, apathetic, and nervous, have tremors, essentially lots of neurological symptoms associated with depression.

"Just a few months previous to the onset of my depression, I had been hospitalized for chest pain, cardiac dysrhysthmia and an inability to take in more than about 1/5 my normal breath. The hospital found no cardiac problems, and the internist gave me an IV drip of magnesium sulfate solution. *A few hours later all of those symptoms vanished as rapidly as they had come. What I was beginning to see was that nearly all illnesses in my adult life were magnesium deficit related.*

"So finally I made the decision to start taking magnesium at a level three times the 400 mg/day RDA for magnesium, with 400 mg in the morning, 400 mg mid afternoon and 400 mg at bedtime. I used Carlson's chelated magnesium glycinate (200 mg magnesium elemental) product. Within a few days to a short week, I felt remarkably better, my depression lifted noticeably, but I was getting a bit of diarrhea.

"Within a week to ten days of starting magnesium, I felt close to being well. I looked so well, that my psychiatrist thought I looked better than he had ever seen me. As I improved, I lowered my dosage of magnesium to find the best dosage for me. I lowered it too much and symptoms rapidly came back. Eventually, I stabilized the dosage at four 200-mg elemental magnesium (as magnesium glycinate) tablets a day. My depression is completely, totally, absolutely gone. I am active and can function mentally, emotionally, and physically at my best again. My vision and bowels also returned to normal."

For all the talk about protecting children in America, too many of our little ones are threatened by psychiatrists and psychologists who have betrayed the young. Millions of children are now taking psychotropic drugs, which are causing catastrophic problems that are going unreported. For a tragic trip into the violent hell these drugs are causing please read Evelyn Pringle's *FDA Forgot A Few ADHD Drug Related Deaths and Injuries at* http://usa.mediamonitors.net/ content/view/full/27099*)*

After High School shootings

The medical and educational establishments are conducting a skyrocketing campaign to get kids and their parents to "just say yes" to brain-altering pharmaceuticals, with the drug of choice being Ritalin, even though some report that Ritalin is a drug that has a more potent effect on the brain than cocaine.[30]

By far, the overwhelming majority of psychotropic prescriptions for children are given for attention deficit disorder (ADD) or attention deficit hyperactivity disorder (ADHD). In some instances, taking medicine is a prerequisite for attending school, with refusal to comply considered grounds for dismissal, or worse, removal of the child from the home by the state.

On top of everything else, the Children's Hospital of Philadelphia has reported that 19% of newly diagnosed Type II diabetic children *also have neurological diseases*. Many of these children are being treated with psychiatric medications Zyprexa, Risperdal, Geodon, Seroquel, Clozaril, and Abilify. All of these drugs carry black box warnings to alert MD's about the dangers of diabetes. All these drugs would in all likelihood push down magnesium serum levels.

Do not, and I scream, do not trust psychologists,
psychiatrists and the current drug-pushing
culture of education.
                                    Dr. Julian Whitaker

There is an international explosion of legal child drugging as parents, educators and politicians en masse have been thoroughly duped into believing that only by continuous heavy drugging from a very early age can the "afflicted" child possibly make it through life's worst challenges.

Teen suicides have tripled since 1960 in the United States. Today suicide is the leading cause of death (after car accidents) for 15 to 24-year-olds. Since the early 1990's millions of children around the world have taken antidepressants that health authorities are just now branding as suicidal agents. This is the other side of the magnesium deficiency, the nightmare of these drugs which only compounds and worsens the loss of magnesium from the body.

The scene has been long in the making for the patterned onslaught of psychiatry on the young. Psychiatry has only in the last two decades unleashed its devastating attack on children using lucrative chemical weapons on — addictive psychotropic drugs posing as medication. Psychiatrists have created a generation of drug addicts and to a great extent they are making the crisis in children today worse when they should be helping to make things better for them.

Child psychiatrists are one of the most dangerous enemies, not only of children but also of adults. They must be abolished.

Dr. Thomas Szasz

Professor of Psychiatry

According to Dr. Sydney Walker, author of *The Hyperactivity Hoax*, "Thousands of children put on psychiatric drugs are simply smart. These students are bored to tears, and people who are bored fidget, wiggle, scratch, stretch, and (especially if they are boys) start looking for ways to get into trouble." What this chapter adds to that is the underlying complication these children face when their magnesium levels are too low and the devastation that rains down on them in the form of psychiatric medications.

If we look at the whole picture what do we see? Children are born under medical stress with unnatural procedures and drugs, they are then vaccinated, i.e., bombarded with chemicals, dosed out with antibiotics, eat nutritionally deficient food, watch inordinate amounts of television, suffer through educational curriculums that, to them, have no relevance in their lives, undergo exposure to thousands of chemical poisons in the environment and home, get more vaccines, stuff their faces to the point of obesity while suffering from malnutrition, only to have to suffer through being drugged by psychiatrists further for becoming the mess they have become.

Psychologists and psychiatrists should know better because of their training in mind and emotion. It is a terrible betrayal of humanity to see them turn into drug dealers. The pharmaceuticals that the drug companies produce for these mental health care workers are as dangerous as any of the drugs dealers on the streets sell. Magnesium should be substituted for these drugs not only because

it is very effective in relieving neurological disorders but because
it is vastly safer than *any* pharmaceutical.

I practiced neurology and psychiatry for 30 years,
but found to my chagrin that it was largely a
huge fraud, despite the fact that most of the doctors
I met had the best intentions. They were simply
brain-washed.

Dr. Alan Greenberg

References

1   Source: www.nytimes.com/2006/02/12/national/12homicide.html?_r=2&th&emc=
th&oref=slogin&oref=slogin

2   Murck H. Magnesium and Affective Disorders. *Nutr Neurosci.*, 2002;5:375-389:
Murck showed many actions of magnesium ions supporting their possible therapeutic
potential in affective disorders. Examinations of the sleep-electroencephalogram
(EEG) and of endocrine system points to the involvement of the limbic-hypothalamus-
pituitary-adrenocortical axis because magnesium affects all elements of this system.
Magnesium has the property to suppress hippocampal kindling, to reduce the release
of adrenocorticotrophic hormone (ACTH) and to affect adrenocortical sensitivity
to ACTH. The role of magnesium in the central nervous system could be mediated
via the N-methyl-D-aspartate-antagonistic, g-aminobutyric acid A-agonistic or the
angiotensin II-antagonistic property of this ion. A direct impact of magnesium on
the function of the transport protein p-glycoprotein at the level of the blood-brain
barrier has also been demonstrated, possibly influencing the access of corticosteroids
to the brain. Furthermore, magnesium dampens the calcium ion-protein kinase C
related neurotransmission and stimulates the Na-K-ATPase. All these systems have
been reported to be involved in the pathophysiology of depression. Murck et al. also
demonstrated induced magnesium deficiency in mice to produce depression-like
behavior which was beneficially influenced with antidepressants.

3   Source: Evelyn Pringle: www.lawyersandsettlements.com/articles/ssri_offlabel.
html

4   Izenwasser SE et al. Stimulant-like effects of magnesium on aggression in mice.
Pharmacol Biochem Behav 25(6):1195-9, 1986.

5   Henrotte JG. Type A behavior and magnesium metabolism. *Magnesium* 5:201-10,
1986.

6   Bennett CPW, McEwen LM, McEwen HC, Rose EL. The Shipley Project: treating
food allergy to prevent criminal behaviour in community settings. *J Nutr Environ Med*

8:77-83, 1998.

7   Kirow GK, Birch NJ, Steadman P, Ramsey RG. Plasma magnesium levels in a population of psychiatric patients: correlation with symptoms. *Neuropsychobiology* 30(2-3):73-8, 1994.

8   Kantak KM. Magnesium deficiency alters aggressive behavior and catecholamine function. Behav Neurosci 102(2):304-11, 1988

9   Buist RA. Anxiety neurosis: The lactate connection. *Int Clin Nutr* Rev 5:1-4, 1985.

10   Seelig MS, Berger AR, Spieholz N. Latent tetany and anxiety, marginal Mg deficit, and normocalcemia. *Dis Nerv Syst* 36:461-5, 1975.

11   Durlach J, Durlach V, Bac P, et al. Magnesium and therapeutics. *Magnes Res* 7(3/4):313-28, 1994.

12   Durlach J. Clinical aspects of chronic magnesium deficiency, in MS Seelig, Ed. Magnesium in Health and Disease. New York, Spectrum Publications, 1980.

13   Kozielec T, Starobrat-Hermelin B. Assessment of magnesium levels in children with attention deficit hyperactivity disorder (ADHD). *Magnes Res* 10(2):143-8, 1997.

14   Kozielec T, Starobrat-Hermelin B. Assessment of magnesium levels in children with attention deficit hyperactivity disorder (ADHD). *Magnes Res* 10(2):143-8, 1997

15   Starobrat-Hermelin B, Kozielec T. The effects of magnesium physiological supplementation on hyperactivity in children with attention deficit hyperactivity disorder (ADHD). Positive response to magnesium oral loading test. *Magnes Res* 10(2):149-56, 1997

16   George MS, Rosenstein D, Rubinow DR, et al. CSF magnesium in affective disorder: lack of correlation with clinical course of treatment. *Psychiatry Res* 51(2):139-46, 1994.

17   Kirov GK, Birch NJ, Steadman P, Ramsey RG. Plasma magnesium levels in a population of psychiatric patients: correlations with symptoms. *Neuropsychobiology* 1994;30(2-3):73-8, 1994.

18   Linder J et al. Calcium and magnesium concentrations in affective disorder: Difference between plasma and serum in relation to symptoms. *Acta Psychiatr Scand* 80:527-37, 1989

19   Frazer A et al. Plasma and erythrocyte electrolytes in affective disorders. *J Affect Disord* 5(2):103-13, 1983.

20   Bjorum N. Electrolytes in blood in endogenous depression. *Acta Psychiatr Scand* 48:59-68, 1972

21   Cade JFJA. A significant elevation of plasma magnesium levels in schizophrenia and depressive states. *Med J Aust* 1:195-6, 1964.

22   Levine J, Rapoport A, Mashiah M, Dolev E. Serum and cerebrospinal levels of calcium and magnesium in acute versus remitted schizophrenic patients. *Neuropsychobiology* 33(4):169-72, 1996.

23   Kanofsky JD et al. Is iatrogenic hypomagnesemia common in schizophrenia? Abstract. *J Am Coll Nutr* 10(5):537, 1991.

24   Kirov GK, Tsachev KN. Magnesium, schizophrenia and manic-depressive disease. *Neuropsychobiology* 23(2):79-81, 1990.

25   Chhatre SM et al. Serum magnesium levels in schizophrenia. *Ind J Med Sci* 39(11):259-61, 1985.

26   Banki CM, Vojnik M, Papp Z, Balla KZ, Arato M. Cerebrospinal fluid magnesium and calcium related to amine metabolites, diagnosis, and suicide attempts. *Biol Psychiatry*. 1985 Feb;20(2):163-71.

27   The effects of magnesium physiological supplementation on hyperactivity in children with ADHD. *Mag Res* 1997; 10(2):149-56.

28   Eli Lilly's newest antidepressant, Cymbalta (Duloxetine) had 6 suicides in the clinical trials, before it ever reached the market, and in people with no previous history of depression. The last and most publicized was the death of a young college girl, who had entered the clinical trial for some extra money while she was in school. She had no depression, was a good student, social and well adjusted, found hanging by a noose after a dosage change of this forecasted blockbuster. See: news.independent.co.uk/uk/health_medical/story.jsp?story=648010

Eli Lilly posted the results of 45 clinical trials on their website in cooperation with recommendations for more transparency for the public, but failed to list the 5 trials with the information about these suicidal acts. Lilly defended its drug, saying that 4,142 depressed patients had taken Cymbalta and the deaths represent a 0.097% suicide rate. Besides, it said, it is the underlying depression — not the drug — that causes sufferers to become suicidal. FDA defended Lilly's position. Later a higher than expected rate of suicide attempts was observed in the open-label extensions of controlled studies of Cymbalta for stress urinary incontinence (SUI) in adult women.

"I have lost my appetite (I weigh around 110 lbs) and loss of taste, hearing, sensation and even judgement. I can barely drive or walk. I am a highly educated woman who has lead a national biotechnology company and now I can barely have a conversation that makes sense. I feel absolutely 100% confident that Cymbalta has caused these side effects. I have a bachelors in science with major studies in pharmacology and medical sciences. I previously had meeting with executives and lead conferences that people would learn from me. Now I can not even stand up straight without feeling dizzy, confused, and paranoid. I am currently having problems with spelling and grammar. This use to be a strong skill of mine. I have been known to edit papers and rewrite many documents and now it is difficult for me to write an email." Cymbalta is not approved for the treatment of SUI. The FDA is evaluating additional data to determine the relationship, if any, between suicidality and Cymbalta use. See: www.fda.gov/cdr/drug/InfoSheets/HCP/duloxetineHCP.pdf

29   Source: www.mdheal.org/magnesiu1.htm

30   West, Jean, "Children's drug is more potent than cocaine," *The Observer*, London, Sept. 9, 2001.

13

||||||||||||||||||||||||||||||||||||||||||||||||||||||||||||||||

Magnesium Chloride

In addition to its immune-stimulating properties,
both magnesium and chloride have other important functions
in keeping us young and healthy. Chloride, is required to
produce a large quantity of gastric acid each day and is also
needed to stimulate starch-digesting enzymes.

There are many compounds that have ionic bonds. They are called ionic compounds, and they are formed when metals react with nonmetals. The formation of magnesium chloride can be thought of as a result from a reaction involving magnesium metal (Mg) and chlorine gas, Cl_2. The reaction involves the following simultaneous processes:

1. The Oxidation of Magnesium Metal. A magnesium atom loses its 2 outer-shell electrons to become a magnesium ion, (i.e. cation). The magnesium metal is said to be oxidized.

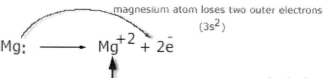

magnesium atom loses two outer electrons $(3s^2)$

$$Mg: \longrightarrow Mg^{+2} + 2\bar{e}$$

magnesium ion has a complete octet of electrons ($1s^2 2s^2 2p^6$)

2. The Reduction of Chlorine Gas. The covalently bonded Cl_2 molecule gains two electrons to become two chloride ions, (i.e. anions). Chlorine is said to be reduced.

3.

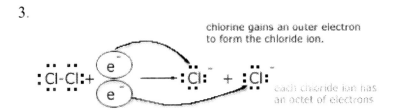

chlorine gains an outer electron to form the chloride ion.

each chloride ion has an octet of electrons

4. Combining the above oxidation and reduction processes, the overall effect is the transfer of TWO electrons from magnesium to chlorine.

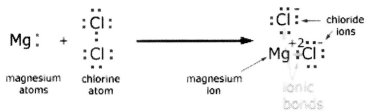

chloride ions

magnesium   chlorine   magnesium   ionic
atoms   atom   ion   bonds

The 2 electrons lost by a magnesium atom are gained by chlorine atoms to produce a magnesium ion and 2 chloride ions.

The opposite charges of the magnesium and chloride ions attract each other, allowing ionic bonds to form. In the solid state, each cation is surrounded by anions, and each anion is surrounded by cations. The ratio of $Mg^{+2}Cl^{-}$ is 1:2. The formula for this ionic compound is $MgCl_2$. As an analogy, the removal of an electron (e -) to a neutral sodium atom (Na) forms a singly charged sodium cation (Na +). Similarly, the removal of two electrons from a neutral magnesium atom (Mg) forms a doubly charged cation (Mg^{+2}).

The "magnesium oil" ($MgCl_2$) (different from $6H_2O\text{-}MgCl_2$ which is the form of crystal or powdered magnesium chloride) is made of one atom of magnesium (atomic weight 24.3) and two atoms of chloride (atomic weight of each is 35.4), which means if a solution contains 35% $MgCl_2$, the percentage of actual magnesium metal ions would be 7.7%. This is about double what you would get as far as magnesium concentration if you use powder or crystal form of magnesium chloride.

Magnesium Chloride is a naturally occurring element and is extracted from salt water solutions such as those found in sea water, Great Salt Lake, Dead Sea and many other locations. To extract the magnesium chloride brine, water is removed from the salt water by solar evaporation. Magnesium chloride has many uses and touches most of us in one way or another every day. In its various forms and concentrations, clean, high quality magnesium chloride has found its way into the food and health business, medical field, industrial uses, agriculture and water operations just to name a few. Magnesium chloride is environmentally safe, and is used around vegetation and in agriculture for dust control.

Magnesium chloride solution is not only harmless for tissues, but it had also a great effect over leucocytic activity and phagocytosis; so it was perfect for external wounds treatment.

Chloride is required to produce a large quantity of gastric acid each day and is also needed to stimulate starch-digesting enzymes. Using other magnesium salts is less advantageous because these have to be converted into chlorides in the body anyway. We may use magnesium as oxide or carbonate but then we need to produce additional hydrochloric acid to absorb them. Many aging individuals, especially with chronic diseases who desperately need more magnesium cannot produce sufficient hydrochloric acid and then cannot absorb the oxide or carbonate.

General Uses for Magnesium Chloride

For the Skin: Sprayed on sun damaged skin regularly will begin to rejuvenate from the inside out and after a few months will be significantly restored. Helps with wrinkles as well as hair health and growth.

Dental: As a mouthwash sprayed into the mouth it is excellent for the gums creating a highly alkaline oral environment. Strengthens teeth and is excellent for gingivitis. It is magnesium, not calcium, which helps form hard tooth enamel resistant to decay.

Mucus Membranes: People have used low concentrations for nose and eye washes and even for application in the vagina. (Beware of stimulative effect).

General Tonic: Magnesium chloride is a strong tonic boosting all aspects of cell physiology and energy production. Magnesium is essential for life as it participates in over 325 enzyme reactions. Expect more energy, strength and endurance and even increased sexual energy.

Sports: Magnesium is perhaps the single most important mineral to sports nutrition. This is true especially of athletes. During vigorous exercise, we excrete critical minerals through our sweat, the most important being magnesium. Adequate magnesium levels will help your body against fatigue, heat exhaustion, blood sugar control and metabolism.

Pain Relief: Transdermal magnesium chloride treatments are essential in the treatment of sport injuries and the aches and pains of sore muscles and joints.

Natural Immune System Booster: Dr. A. Neveu observed that magnesium chloride has no direct effect on bacteria (i.e. it is not an antibiotic). Thus he thought that its action was a specific, immunity enhancer, so it could be useful against viral diseases.

Memory and Cognitive Function: Magnesium deficit may lead to decreased memory and learning ability, while an abundance of magnesium may improve cognitive function in children and the elderly.

Blood Brain Barrier

Several studies show that magnesium does cross the blood–brain barrier, in both animals and in humans.[1] Brain magnesium concentrations are regulated by active blood–brain barrier transport.[2] Cerebrospinal fluid magnesium concentration increases by 20-25% in response to doubling of the serum concentration, and peaks around four hours after parenteral administration.[3]

The concept of a barrier between the blood and brain interface is about one hundred years old. The brain barriers, namely blood-brain barrier (BBB) and blood-cerebrospinal fluid barrier, usually referred to as the choroid plexus maintains the chemical stability of the central nervous system (CNS). Less known are the functions of the BBB in brain development, neuroendocrine regulation, drug efflux and metabolism, as well as aging processes. The choroid plexus embraces the cerebrospinal fluid (CSF) compartment, the interstitial fluid (ISF) or extracellular fluid compartment, and the intracellular compartment playing a pivotal role in maintaining the homeostasis of essential metal ions[4] in the CNS.

The composition of the cerebrospinal fluid with a specific gravity ($1.004 - 1.007$ g/cm^3) is much like blood plasma; it is a clear, colorless fluid that contains glucose, proteins, lactic acid, urea, salts, and some white blood cells. The cerebrospinal fluid picks up metabolic wastes as it circulates past the nervous tissue of the brain and spinal cord. These metabolic wastes then move into the bloodstream in the intracranial vascular sinuses as the CSF is absorbed. The blood carries these wastes away to be eliminated from the body by the lungs and kidneys.[5] Changes in the composition (increased protein) or in the appearance (cloudiness) of the CSF would suggest some neurologic disease.[6]

Systemic Zinc deficiency can also affect the permeability of brain barriers. Zinc deficiency significantly increases the permeability of the blood–brain barrier.

Noseworthy and Bray (2000)

The choroid plexus separates the CSF compartment from the systemic blood compartment and possesses numerous transporters for metals, metal–amino acid conjugates, and metal–protein complexes. There are many hundreds of different transporter types, each specialized for different substances. The integrity and function of the BBB is mission critical for overall brain function. Changes in permeability often reflect alterations in BBB transport systems. Toxicological causes of generalized changes in BBB permeability include organic solvents, enzymes, and heavy metals. Some agents like mercury induce selective changes in BBB transport at very low doses.

Brain barrier integrity is compromised by free radicals.

Magnesium has been seen to attenuate increased blood-brain barrier permeability during insulin-induced hypoglycemia in animal studies. Magnesium has its important role at the BBB and researchers think that this metal probably protects brain tissue against the effects of cerebral ischemia, brain injury, and stroke through its actions as a calcium antagonist and inhibitor of excitatory amino acids.

Children are most susceptible to brain damage
because the blood/brain barrier has not had time
to develop enough to filter out poisonous substances
like lead and mercury and the other heavy
metals, drugs and chemicals that
are assaulting their systems.

As amino acids and their associated linkage are highly susceptible to enzymatic degradation, the nature and concentration of specific enzymes at the BBB can greatly impact the efficacy of detoxification and nutrient supply. Magnesium is crucial in preventing enzyme degradation and thus crucial for BBB integrity. Specific transporters exist at the BBB that permit nutrients to enter the brain and toxicants / waste products to exit. Independent transport systems for glucose, neutral amino acids, basic amino acids, and monocarboxylic acids have been identified in the BBB.

Especially relevant is the transport of methyl mercury into the brain. In vivo studies in rats have demonstrated that methyl mercury, bonded to the amino acid cysteine, is transported across the blood-brain barrier. Lead transport at the blood-brain barrier is dependent on the ATP calcium pump and this pump is dependent on magnesium. For the lead to get out of the brain, the pump must be working properly.[7]

Excitotoxicity, a mechanism by which excess glutamate accumulates outside the neuron, thereby leading to death of the cell by an excitation process, has been linked to mercury neurotoxicity. Recent studies have confirmed that mercury, even in concentrations below that known to cause cell injury, paralyzes the glutamate removal mechanism, leading to significant damage to synapses, dendrites and neurons themselves. Glutamate and its biochemical "cousin," aspartic acid or aspartate, are the two most plentiful amino acids in the brain.

*Wheat gluten is 43% glutamate, the milk
protein casein is 23% glutamate.*

This glutamate removal mechanism is critical to brain protection. Additionally, mercury in very low concentrations increases glutamate release, primarily by stimulating the brain's immune cell, the microglia. Chronic microglial activation, as seen with mercury exposure, has been linked to neurodegenerative diseases. Mercury, among all the metals tested, was the only one shown to block the removal of excess glutamate from the nervous system. By paralyzing the glutamate removal system, mercury triggers chronic excitotoxicity — that is chronic destruction of the nervous system.

*Excess glutamate can also produce the same
neurofibrillary tangles seen with mercury exposure.*

Glutamate transport at the BBB is crucial and mercury, if not neutralized, plays havoc at the barrier. Two of the principle conditions that allow glutamate to shift from neurotransmitter to excitotoxin are:

1) Inadequate neuronal ATP levels (whatever the cause)
2) Inadequate neuronal levels of magnesium

One of the most common food additives, MSG (monosodium glutamate), has expanded greatly in use, doubling every decade since 1948. Aspartic acid is one half of the now ubiquitous sweetener aspartame (NutraSweet®), which is the basis of diet desserts, low-calorie drinks, chewing gum, etc. Both of these food additives spell danger for our children.

Glutamate and aspartate are neurotransmitters. Neurotransmitters are the chemicals that allow neurons to communicate with and influence each other. Neurotransmitters serve either to excite neurons into action, or to inhibit them. Glutamate receptors are excitatory — they literally excite the neurons that contain them into electrical and cellular activity. When glutamate or aspartate attaches to the NMDA (N-methyl-D-aspartate) receptor, it triggers a flow of sodium (Na) and calcium (Ca) ions into the neuron, and an outflow of potassium (K). When the PCP is occupied, the opening of the ion channel is blocked, even when glutamate occupies its receptor site. Magnesium can occupy and block the NMDA channel, which means that as long as the neuron is able to maintain its normal resting electrical potential of -90 millivolts, the magnesium blocks the ion channel even with glutamate in its receptor.

Safety

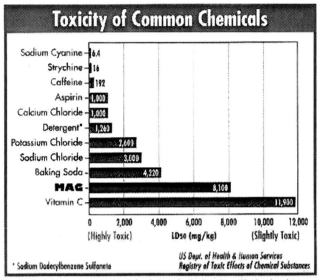

The data for MAG is for magnesium chloride hexahydrate, the data for Calcium Chloride is for calcium chloride anhydrous.

The longer the bar in the above graph the safer the substance.

Many of the best magnesium chloride products are made from condensed seawater. In Japan, ion exchange filters are used that have extremely fine pores 1/100th of a millimeter (.001 microns) small, which allows magnesium, calcium and potassium to pass through, but not large molecules such as PCBs (dioxins) or heavy metals (mercury, arsenic, etc.). The resulting salt water is processed to form salt crystals in a vacuum style vaporization canister.

The best magnesium oils available could appropriately be called low, or even no technology. Using seawater and a long evaporation process of two years, it is one of the least processed products on the market.

Dr. Boyd Haley, the former chairman of the chemistry department of the University of Kentucky, and the world's leading expert on mercury toxicity, ran tests and said, "The magnesium oil is safe. There is a small amount of mercury in it but there is a several fold excess of selenium also. The selenium would react with the mercury, forming a mercury/selenium compound (HgSe), which is one of the most stable and least toxic of all mercury compounds." Dr. Haley found 30 times more selenium than mercury, "which would bind the mercury and render it much, much less toxic."

The IMC Research Laboratory in Cyprus in the Mediterranean, tested the magnesium oil using an ICP-OES with detection limits in the parts per billion and found no detectable mercury. Dr. Haley was using instruments sensitive to parts per trillion so the actual level is just around the four parts per billion mark. Dr. Haley reported, "Our instrument can measure 0.01 nanograms of mercury (that is 10^{-9} grams).

We found the level of mercury in the magnesium oil to be in the range of 4,000 nanograms per liter or 4 micrograms per liter or 0.004 milligrams per liter. A liter of water is one million milligrams so the 0.004 milligrams is about 0.004 ppm or 4 ppb. Now, the amount is very low and sea water is known to also have about 0.12 to 0.15 mg per liter of selenium, which is in excess of the mercury by about three to four-fold by weight. Therefore, it is very unlikely that any free mercury is available to cause toxic problems as it would be tied up in the non-reactive mercury/selenium form."

The total heavy metal profile of the best ocean-derived magnesium chloride adds up to only 416 parts per billion or .0000416%. Most food grade powder magnesium chloride products claim higher heavy metal toxicities (Heavy Metals: 0.001% max) making the magnesium oil one of the safest magnesium chloride products found to date.

MAGNESIUM OIL ANALYSIS

Element	Result (ppb)
Aluminum	248.5
Antimony	<dl
Arsenic	74.05
Beryllium	<dl
Cadmium	11.73
Cobalt	<dl
Lead	82.64
Molybdenum	<dl
Mercury	<dl
Nickel	<dl
Calcium	<dl
Chromium	<dl
Copper	26.31
Iron	163.1
Magnesium	Saturated
Manganese	62.16
Selenium	22.26
Strontium	12.42
Zinc	<dl

Dosage

Before one begins transdermal magnesium chloride it is highly advised that you read the chapter on warnings and contraindications if you are suffering from any chronic illness, severe disease or deficiency, or are taking any pharmaceutical medications. Also if one is suffering from any kind of disease it is always recommended to have your treatments supervised by a primary health care practitioner. That could be a nurse, chiropractor, naturopathic doctor, acupuncturist, or allopathic medical doctor. Unfortunately few know anything about transdermal magnesium mineral therapy because it is so new. The most pertinent question about magnesium chloride dosing is: how effectively is it absorbed transdermally?

Magnesium chloride is without doubt a versatile mineral medicine, though as with all forms of magnesium supplementation, it is not easy to calculate the exact dosage. Absorption rates vary considerably from one person to another and from one form of use to another, even with magnesium chloride, which probably delivers more usable magnesium to the cells than any other form. *It is wise, especially if one is seriously ill, to start out with low dosages and build slowly up to higher doses over a period of a week or two.*

In general, to individualize the appropriate magnesium dosage for oral intake, the rule of thumb is approximately 6-8 mg/kg (3-4 mg per pound) of body weight per day. That translates into a total dietary magnesium intake of 600 to 900 mg per day for a 200 lb. man. With children some researchers indicate that 10 mg/kg/day are appropriate because of their low body weight and increased requirements for growth. Athletes also need more depending on their stress and training levels[8] and we can always adjust upwards when under great emotional stress or when seriously ill.[9,10]

The normal accepted recommended daily dietary amount of magnesium is only 300-400 mg. Many professionals feel this to be the bare minimum. Some would say that 1,000 mg is probably more in the range of what most people need due to stress (measured by cortisol levels) causing magnesium to be dumped into the sweat in increasing quantities. Most people are numb to the amount of stress experienced every day. But cortisol can be measured by saliva tests, if one really wants to know, and if found to be high, magnesium dosages can be adjusted up accordingly.

Dr. Norm Shealy, who has tested the transdermal/topical method against oral and intravenous applications, asserts that only through the transdermal form are DHEA levels raised. According to Shealy the best absorbed oral preparation is magnesium taurate, but in his experience, it takes up to one year of oral supplementation to restore intracellular levels to normal.

Until a few years ago, Dr. Shealy gave most of his patients' ten doses of magnesium chloride intravenously over a period of two weeks. This helped to restore the intracellular levels to normal and usually allowed them then to maintain normal levels with oral supplementation. However, one can use transdermal magnesium mineral therapy to achieve the same result in only a slightly longer time frame. In four weeks, use of magnesium oil can accomplish as much as having the ten doses intravenously according to Shealy who says, "It is a lot simpler and easier, and you can do it on your own. There is no known risk to using magnesium unless you have kidney failure."

As with anything just starting, caution should be taken in the beginning until one gets a feel for the appropriate dosages for adult and especially with children. Each person has to adjust the dosage to their own needs, size and body weight. The actual amount used is also dependent on the method of use or the combinations of methods used. Magnesium chloride may be taken orally, applied directly to the skin (used in a massage or simply rubbed on), used in foot baths, full body baths, and sprayed into mucus membranes.

Our cells are best served when they are brimming with magnesium reserves and we need to absorb a sufficient amount each and every day. A magnesium saturated body will have a tougher immune system that will fight more easily against infections and influenza. This does not mean that we should all put ourselves into hypermagnesemia without concern and ignore the needed balance with other minerals. What we really have to do is make sure we have adequate magnesium, for all the cellular systems to work to their optimal level.

The requirements for a very ill person are going to be higher than for a healthy person. In general, for a large adult, spraying one ounce of magnesium oil a day all over the body is recommended with that adjusted downward for children depending on their age and size. If applied in a full body bath two ounces or more could be used. Some people enjoy a very concentrated magnesium chloride bath applying as many as eight ounces at a time. For sport injuries more concentrated baths would definitely be indicated. Foot baths use much less water so two ounces will yield a very concentrated intake.

The magnesium oil can and should be diluted when applying directly to the skin (especially with children) if redness or "stingy" feelings result in uncomfortable feelings or sensations. If one is suffering from long term illness of any kind, dosages, whether orally or topically administered, should be started at lower levels and brought up gradually. Magnesium chloride and vitamin C have similar toxicity profiles with overdose from both resulting at worst usually in diarrhea unless the kidneys are seriously compromised.

Soak the whole body or just the feet in bath water for 20-30 minutes, at a temperature of about 108° (F). The most effective protocol for this therapy is to begin with a daily body or foot bath every day for the first 7 days, (starting at lighter concentrations and building up) then continue with a maintenance program of 2-3 times a week for 6-8 weeks or longer.

Sensitive care must be taken with children as to dose levels, water temperature and magnesium concentrations. Muscle spasms might occur on rare occasions if one forgets to get out of the tub so it is necessary to supervise children and the length of time they remain soaking in magnesium chloride. All strong reactions like redness in local areas to diarrhea or even muscle spasms are indications to reduce concentration.

Fick's Law of Membrane Permeability says that the amount of any solute (magnesium) that will be absorbed is directly dependent upon the area of contact, the concentration of the solution, and the time that the solute is in contact with the membrane.[11] Thus one has to feel one's way to appropriate dosage both in initial self-treatment phases and for long term maintenance dosage levels.

A particularly strong sensation is realized when one uses magnesium chloride in the mucous membranes and it is especially useful as a mouthwash to strengthen teeth and revitalize the gums. Applying three or four sprays full or half strength twice a day is appropriate.

There are no numbers available for how many milligrams are absorbed through the skin but it is generally acknowledged by those who have been involved with transdermal application of magnesium chloride that it is an important way to supplement magnesium.

Magnesium oil from the sea weighs 12 pounds per gallon. Distilled water weighs only 8 pounds.[12] Thus we can calculate in a straightaway manner how much elemental magnesium is in each gallon and ounce. Each spray of magnesium oil contains approximately 18 milligrams of elemental magnesium. An ounce would contain just over 3,300 mg. Five sprays in a glass of water would be almost 100 milligrams and one could probably count on the majority of that being absorbed.

If two ounces are put into a bath we might have over 6,600 milligrams floating around in the water, although only a fraction of that will be absorbed. But absorbed it will be for almost everyone experiences the effect of deep relaxation. Spraying it on the body will yield a higher magnesium concentration on the skin so an ounce used that way will result in more magnesium absorbed than two ounces used in a bath.

It should be understood that we need more research into studies on absorbability and bioavailability through the skin. Possibly what is best is a combination approach — alternating between baths, direct spraying on the body, and oral intake while increasing the intake of foods high in magnesium. When one uses all avenues together it is easier to bring ones magnesium levels up and then to maintain one's intake.

Magnesium Chloride	1 oz.	3,300mg.
Food Sources of Magnesium		
Spirulina,	1 oz.	110mg
Tofu, firm,	1/2 cup	118mg
Chili with beans,	1 cup	115mg
Wheat germ, toasted,	1/4 cup	90mg
Halibut, baked,	3 oz.	78mg
Swiss Chard, cooked,	1 cup	75mg
Peanut, roasted,	1/4 cup	67mg
Baked potato with skin,	1 medium	55mg
Spinach, fresh,	1 cup	44mg

— Source: USDA: Composition of Foods. USDA Handbook No. 8 Series. Washington, D.C., ARS, USDA, 1976-1986.

There is no specific information about oral magnesium chloride in liquid form but it is reasonably safe to assume it would be more absorbable than magnesium taurate because liquid minerals are in general more absorbable than tablets.

The taste of the solution is not very good (it has a bitter-saltish flavor) so a little fruit juice (grapefruit, orange, lemon) can be added to the solution. Individuals with very sensitive taste buds may start using it in small amounts and increase doses very gradually. Alternatively, drink it in one gulp dissolved in water while pinching your nose and quickly drink something afterwards.

> *3-5 sprays of magnesium chloride in a glass*
> *of pure water is an excellent way to take magnesium*
> *internally. It assists digestion, counteracts excess*
> *acidity in the stomach, and delivers magnesium swiftly into*
> *the bloodstream for distribution to all the cells of the body.*
>
> Daniel Reid
> *The Tao of Detox*

Hydrated magnesium chloride (powder or crystal) contains about 120 mg of magnesium per gram or 600 mg per rounded teaspoon. It has a mildly laxative effect. As a good maintenance intake to remain healthy you may take about 400 mg or a level teaspoon daily in divided doses with meals.

With raised blood pressure and symptoms of magnesium deficiency you may temporarily increase this to 2 teaspoons daily in divided doses under the supervision of your health care practitioner. This may already cause "loose stools" in some. However, commonly with these conditions a rounded teaspoon daily or 600 mg may be just right.

Dr. Raul Vergini offers the following guidelines for oral intake of a 2.5% magnesium chloride hexahydrate ($MgCl_2$-$6H_2O$) solution (i.e.: 25 grams or approximately 1 oz. of pure food grade powder in a liter of water). The quantity of elemental magnesium contained in a 125 cc dose of the 2.5% solution is around 500 mg.

Dosages are as follows:

Adults and children over 5 years old	125 cc
4 year old children	100 cc
3 year old children	80 cc
1-2 year old children	60 cc
Over 6 months old children	30 cc
Under 6 months old children	15 cc

125 milliliter = 4.2267528 ounce [US, liquid]
cc and ml are equivalent

Dr. Vergini indicates that "In acute diseases the dose is administered every 6 hours (every 3 hours the first two doses if the case is serious); then space every 8 hours and then 12 hours as improvement goes on.

After recovery it's better going on with a dose every 12 hours for some days. As a preventive measure, and as a magnesium supplement, one dose a day can be taken indefinitely. Magnesium chloride, even if it's an inorganic salt, is very well absorbed and it's a very good supplemental magnesium source."

Daniel Reid says, "Using magnesium oil is the quickest and most convenient way to transmit magnesium chloride into the cells and tissues through the skin. Two to three sprays under each armpit function as a highly effective deodorant, while at the same time transporting magnesium swiftly through the thin skin into the glands, lymph channels, and bloodstream, for distribution throughout the body. Spray it onto the back of the hand or the top of the feet any time of day or night for continuous magnesium absorption. Regardless of where you apply the spray on the body, once it penetrates the surface of the skin, the body transports it to whichever tissues need magnesium most."

Typical recommendations for magnesium oil made from $MgCl_2$-$6H_2O$ suggest using 6-8 oz in each bath. With ocean-derived magnesium oil, the recommended amount per bath is only 2 oz. This magnesium oil is 30%–35% magnesium chloride as opposed to approximately 25% for the magnesium oils one makes from powder or crystal magnesium chloride hexahydrate. Cost analysis between the products need to reflect the large difference in dosage required.

All massage therapists should be using magnesium oil for it is always a good idea to combine a massage with a magnesium treatment. If we really appreciated how important it is to make sure our magnesium levels are satisfactory we would be spraying our underarms with it everyday, spraying it on to different parts of our body and would never leave it out of our baths.

References

1   Muir KW. Magnesium for neuroprotection in ischaemic stroke: rationale for use and evidence of effectiveness. *CNS Drugs.* 2001;15:921–930.

2   Oppelt WW, MacIntyre I, Rall DP. Magnesium exchange between blood and cerebrospinal fluid. *Am J Physiol.* 1963;205:959–962.

3   Fuchs-Buder T, Tramer MR, Tassonyi E. Cerebrospinal fluid passage of intravenous magnesium sulfate in neurosurgical patients. *J Neurosurg Anesthesiol.* 1997;9:324–328.

4   Research in the last several decades has revealed that at least 11 metals—lead (Pb), mercury (Hg), cadmium (Cd), manganese (Mn), arsenic (As), iron (Fe), copper (Cu), zinc (Zn), silver (Ag), gold (Au), and tellurium (Te)—accumulate in the choroid plexus (Zheng, 2001a, 2002), making the tissue a major target in brain for toxicities associated with environmental exposure to heavy metals.

5   Morehead State University: BIOL 231 Human Anatomy (See http://people. morehead-st.edu/fs/m.mcmurr/231-L25.html)

6   University of Manitoba: Cerebral Ventricular System and Cerebrospinal Fluid. (See www.umanitoba.ca/faculties/medicine/anatomy/csf-form.htm)

7   University of Rochester Environmental Health Sciences Center, Clarkson (See www-apps.niehs.nih.gov/centers/public/res-core/ctr1082-4386.htm)

8   Seelig, MS. Athletic stress, performance and magnesium in consequences of magnesium deficiency on the enhancement of stress reactions; preventive and therapeutic implications:a review. *J Am Coll Nutr*, Vol.13, No. 5, pp. 429-446, 1994

9   Durlach, J. *Magnesium in Clinical Practice*, Libbey, London, 1988.

10   Fehlinger, R. Therapy with magnesium slats in neurological diseases. *Magnes Bull*, Vol. 12, pp. 35-42, 1990

11   Diffusion is the mechanism by which components of a mixture are transported around the mixture by means of random molecular (Brownian) motion (cf. permeation: the ability of a diffusant to pass through a body - dependent on both the diffusion coefficient, D, and the solubility coefficient, S, ie, permeability coefficient, P = D.S). Flynn et al. cite Berthalot as postulating, at the beginning of the nineteenth century, that the flow of mass by diffusion (ie, the flux), across a plane, was proportional to the concentration gradient of the diffusant across that plane. (See www.initium.demon. co.uk/fick.htm)

12   Magnesium chloride is an ionic compound because it has a metal, magnesium, and a nonmetal, chlorine. Magnesium will lose two electrons and form a +2 charge. Chlorine will gain one electron to form a chloride ion with a -1 charge. The formula for the compound is $MgCl_2$. To get the formula weight, find the atomic weights and add them together taking the subscripts into account. Magnesium is 24.3; chlorine is 35.5; so two would be 71.0. The total gives 95.3 as the formula weight.

14

||||||||||||||||||||||||||||||||||||||||||||||||||||||||||||||||||||||||||

Magnesium Chloride vs. Magnesium Sulfate

According to Daniel Reid, author of *The Tao of Detox*, magnesium sulfate, commonly known as Epsom salts, is rapidly excreted through the kidneys and therefore difficult to assimilate. This would explain in part why the effects from Epsom salt baths do not last long and why you need more magnesium sulfate in a bath than magnesium chloride to get similar results. Magnesium chloride is easily assimilated and metabolized in the human body.[1] Epsom salts are used by parents of children with autism because of the sulfate, which they are sometimes deficient in; sulfate is also crucial to the body and is wasted in the urine of autistic children.

Dr. Jean Durlach et al, at the Université P. et M. Curie, Paris, wrote a paper about the relative toxicities between magnesium sulfate and magnesium chloride. They write, "The reason of the toxicity of pharmacological doses of magnesium using the sulfate anion rather than the chloride anion may perhaps arise from the respective chemical structures of both the two magnesium salts.

Chemically, both magnesium sulfate ($MgSO_4$) and magnesium chloride ($MgCl_2$) are hexa-aqueous complexes. However $MgCl_2$ crystals consist of dianions with magnesium coordinated to the six water molecules as a complex, $[Mg(H_2O)6]2+$ and two independent chloride anions, Cl^-. In $MgSO_4$, a seventh water molecule is associated with the sulphate anion, $[Mg(H_2O)6]2 +[SO_4. H_2O]$. Consequently, the more hydrated $MgSO_4$ molecule may have chemical interactions with paracellular components, rather than with cellular components, presumably potentiating toxic manifestations while reducing therapeutic effect."

> *$MgSO_4$ is not always the appropriate salt in clinical therapeutics. $MgCl_2$ seems the better anion-cation association to be used in many clinical and pharmacological indications.[2]*
> Dr. Jean Durlach et al

These researches also studied ionic fluxes in the two directions between the mother and the fetus. They found that there was a greater positive effect when $MgCl_2$ was used and that $MgSO_4$ could not guarantee the fetal needs in sodium and potassium exchange like $MgCl_2$ could. Dr. Durlach summarized saying, "$MgCl_2$ interacts with all exchangers while the interaction of $MgSO_4$ is limited to paracellular exchangers, and $MgCl_2$ increases the flux ratio between mother to fetus while $MgSO_4$ decreases it."

Magnesium Effects During Pregnancy

Dropping levels of magnesium during pregnancy leads to premature contraction and this has been treated by allopathic medicine mostly with magnesium sulfate. But high-dosage, tocolytic magnesium sulfate administered to pregnant women during pre-term labor can be toxic, and sometimes lethal, for their newborns.[3]

A Medline search found $MgSO_4$ had 53 reports of its use in prematures,[4] whereas $MgCl_2$ had only 4 papers of its use. The paper sited just above showed the results of severe overdose of the mothers, 50 grams or more of $MgSO_4$. Clearly, too much is toxic, but other studies show safety and efficacy at lower doses. Magnesium sulfate given to women immediately before very pre-term birth may improve important pediatric outcomes. No serious harmful effects have been seen at lower dosage levels.

Chloride is required to produce a large quantity of gastric acid each day and is also needed to stimulate starch-digesting enzymes. We may use magnesium as oxide or carbonate but then we need to produce additional hydrochloric acid to absorb them. Many aging individuals, especially with chronic diseases who desperately need more magnesium cannot produce sufficient hydrochloric acid and then cannot absorb the oxide or carbonate.

Sulfate is also important and has an influence over almost every cellular function. Sulfate attaches to phenols and makes them less harmful, and sets them up for being excreted from your kidneys. A lot of these potentially toxic molecules are in food. Sulfate is also used to regulate the performance of many her molecules. Many systems in the body will not function well in a low-sulfate environment. Sulfur is so critical to life that the body will apparently borrow protein from the muscles to keep from running too low.

Though magnesium sulfate will save your life in emergency situations as quickly and easily as magnesium chloride, magnesium chloride fits the bill as a universal medicine, magnesium sulfate does not. Magnesium sulfate is a close cousin whose effect, form and toxicity demands it be used in special applications when the sulfur is needed.

References

1   See: www.hps-online.com/foodprof14.htm

2   *Magnesium Research*. Volume 18, Number 3, 187-92, September 2005, original article

3   Mittendorf R, Dammann O, Lee KS. Brain lesions in newborns exposed to high-dose magnesium sulfate during preterm labor. Department of Obstetrics and Gynecology, Loyola University Medical Center, Maywood, IL, USA. *J Perinatol*. 2005 Dec 1; doi:10.1038/sj.jp.7211419.

4   Dr.Herbert C. Mansmann Jr. and his associates show us the incredible importance of supplementing magnesium in pregnancy. "The survival rate of very preterm, low birth weight infants (weighing less than 1500 g) is 85 per cent in the USA and is ever increasing, while 42 to 75 per cent of extremely premature infants (weighing 751-1000 g) survive. Of great concern is the lack of consistent decrease in neurological syndromes and associated visual impairments. Because of short gestations, these infants have not had time to accrue up to 80 per cent of magnesium normally present at term. These very preterm infants are at highest risk for cerebral hypoxia/ischemia (H/I), intracranial hemorrhage (ICH), periventricular leukomalacia (PVL) or cystic PVL (CPVL), and possible sequelae, cerebral palsy (CP) and mental retardation (MR)." Department of Pediatrics, Thomas Jefferson University, Philadelphia, PA 19107-5083, USA. The possible role of magnesium in protection of premature infants from neurological syndromes and visual impairments and a review of survival of magnesium-exposed premature infants. *Magnesium Research*. 12(3):201-16, 1999 Sep.

15

|||||||||||||||||||||||||||||||||||||||||||||||||||||||||||||||||||||||

Memory, Cognitive Function, and Periodontal Disease

Magnesium may reverse middle-age memory loss.[1]
Massachusetts Institute of Technology

Associate Professor Guosong Liu and postdoctoral associate Inna Slutsky at MIT's Picower Center for Learning and Memory found that magnesium helps regulate a key brain receptor important for learning and memory. Their work provides evidence that a magnesium deficit may lead to decreased memory and learning ability, while an abundance of magnesium may improve cognitive function.

"Our study shows... maintaining proper magnesium in the cerebrospinal fluid is essential for maintaining the plasticity of synapses," the authors wrote. "Since it is estimated that the majority of American adults consume less than the estimated average requirement of magnesium, it is possible that such a deficit may have detrimental effects... resulting in potential declines in memory function."

Plasticity, or the ability to change, is the key to the brain's ability to learn and remember. Synapses, the connections among brain cells, undergo physical changes in response to brain activity. While the mechanisms underlying these changes remain elusive, it is known that synapses are less plastic in the aging or diseased brain. Loss of plasticity in the hippocampus, where short-term memories are stored, causes the forgetfulness common in older people.

Armed with this new understanding, the researchers then identified magnesium's importance in synaptic function. Magnesium is the gatekeeper for the NMDA receptor, which receives signals from an important excitatory neurotransmitter involved in synaptic plasticity. Magnesium helps the receptor open up for meaningful input and shut down to background noise. "As predicted by our theory, increasing the concentration of magnesium and reducing the background level of noise led to the largest increases of plasticity ever reported in scientific literature," Liu said.

The researchers have identified and are now studying several families of drugs that may restore learning and memory in animals. Most important, Liu said, "This new theory may help create strategies to prevent aging-induced loss of synaptic plasticity."

Magnesium Deficiency & Periodontal Disease

Many chronically ill patients have periodontal problems. An association between magnesium and periodontitis has been suggested. In a study, conducted by the International and American Associations for *Dental Research*, subjects aged 40 years and older, increased serum magnesium/calcium was significantly associated with reduced probing depth ($p < 0.001$), less attachment loss ($p = 0.006$), and a higher number of remaining teeth ($p = 0.005$). Subjects taking magnesium showed less attachment loss ($p < 0.01$) and more remaining teeth than did their matched counterparts.[2] These results suggest that increased magnesium supplementation will improve periodontal health.

> *It is magnesium, not calcium, that helps*
> *form hard tooth enamel, resistant to decay.*

I use magnesium chloride as a mouthwash for my gum problems and felt a change in my oral environment after only one application. I had periodontal disease when I was seventeen and severe bone loss through many years. My mouth was also loaded up with mercury amalgam starting at age five. It is very difficult to keep up the discipline of maintaing my mouth perfectly. A fine dentist told me six years ago that I had to brush and floss after every meal and never eat between meals to preserve my teeth.

I have had a hard time following those directions and the results are as expected. I now use salt water instead of fluoride toothpaste and that has helped out a lot. Spraying in three pumps of the magnesium chloride though has had a dramatic healing and alkalizing effect. My whole mouth feels stronger and my oral environment stays with a healthier feeling for hours more if I do not clean my teeth.

References

1   See: web.mit.edu/newsoffice/2004/magnesium.html

2   P. Meisel 1 et all. Magnesium Deficiency is Associated with Periodontal Disease *Dent Res* 84 (10):937-941, 2005 International and American Associations for Dental Research

16

||||||||||||||||||||||||||||||||||||||||||||||||||||||||||||||||||||

Natural Influenza Protocol with Magnesium and Vitamin C

In the winter of 2005, The New York Times wrote, "As concern about a flu pandemic sweeps official Washington, Congress and the Bush administration are considering spending billions to buy the influenza drug *Tamiflu*. But after months of delay, the United States will now have to wait in line to get the pills." The time of year had come when the Centers for Disease Control (CDC), like kids eager to light the family Christmas tree, put the plug back in the wall with the intention of lighting up signs all over the country. However, instead of spreading joyful, reassuring news, they would instead whip the public into its annual flu frenzy.

Not only did we have to worry about the regular flu and its misery, but now we had the bird flu. On October 5, 2005 the New York Times reported that the deadly 1918 influenza pandemic was linked to the avian flu. Two teams of federal and university scientists announced that they had resurrected the 1918 influenza virus and found that it was actually a bird flu that jumped directly to humans.

Democrats on Capitol Hill lodged their partisan complaints that the delay put Americans in jeopardy. Tamiflu, introduced in 1999, had become the drug of choice of the medical establishment worried about pandemic flu, because it was one of the only medicines claimed to reduce the duration and severity of the potentially deadly disease if taken within 48 hours of infection.

Tamiflu was "supposed" to speed recovery from the flu. When started during the first two days of the illness, it was said to hasten improvement by at least "a day."

Tamiflu belongs to a class of antiviral drugs called neuraminidase inhibitors. No studies had been conducted to compare Tamiflu with mother's chicken soup, which is also reported to diminish intensity, discomfort and duration of the flu. However, the efficacy of the CDC's claim was taken as the gospel truth. This creates a demand for the drug, much to the delight of its manufacturer and its shareholders, but also creates a shortage, and panic, all of which are stress inducers, which further compromises the health of those who believe the drug is their *only* protection and salvation. Most importantly, it does not necessarily lower the risk of, or actually prevent the disease.

This is one of many examples of how medical officials line up to scare the wits out of the public, only to offer ineffective and impotent medical strategies.

"This is a nation-busting event!" warned Dr. Tara O'Toole of the University of Pittsburgh Medical Center's Center for Biosecurity, commenting on the avian flu threat. Speculating that

40 million Americans could die — that's about one in eight — she warned: "We must act now!"

For the avian flu medical authorities are rushing to develop new unproven vaccines. Concerned that a resistance to Tamiflu's effects was growing, other drugs like Relenza were recommended for stockpiling.[1]

"We and the entire world remain unprepared for what could arguably be the most horrific disaster in modern history," said Dr. Gregory A. Poland of the Mayo Clinic speaking about the avian flu. "The key to our survival, in my opinion, and to the continuity of government is vaccination." And "we do not have a licensed or approved vaccine," continued Dr. Poland.

Dr. J. Anthony Morris, former Chief Vaccine Control Officer for the FDA said, "There is no evidence that any influenza vaccine thus far developed is effective in preventing or mitigating any attack of influenza. The producers of these vaccines know that they are worthless, but they go on selling them, anyway."

> *A review in* The Lancet *suggests that*
> *influenza vaccination of infants is useless.*
> Dr. F. Edward Yazbak

Each and every year at this time medical officials come out to remind us of our mortal danger. However, they advocate medical procedures that do little to nothing to protect us, but cost billions nonetheless.

Dr. Eleanor McBean was an on-the-spot observer of the 1918 Influenza epidemic and said, "As far as I could find out, the flu hit only the vaccinated. Those who had refused the shots escaped the flu. My family had refused all the vaccinations so we remained well all the time. We (who didn't take any vaccines) seemed to be the only family which didn't get the flu. It has been said that the 1918 flu epidemic killed 20,000,000 people throughout the world. But, actually, the doctors killed them with their crude and deadly treatments and drugs. This is a harsh accusation but it is nevertheless true."

The Vaccine Injury Compensation Program created a federal, court-like process through which victims of vaccination could seek financial compensation. Created in 1986, the program has settled 1,200 vaccine claims worth $1.2 billion as of 2004. This represents compensation to only a fraction of the children and families devastated by vaccines.

Most problems noted during tests of Tamiflu were indistinguishable from the symptoms of flu. That's a tricky way of saying Taniflu can as easily cause the flu as diminish it. Tamiflu should not in fact be the drug of choice because of its large side-effect profile that leaves one wondering if their flu symptoms are from a virus or Tamiflu. Below is the list of side-effects that Tamiflu is known to cause. These are all the reasons Tamiflu should be avoided:

Aches and pains	Indigestion
Allergic reactions sometimes leading to shock	Liver problems
Asthma — aggravation of pre-existing asthma	Lymphadenopathy
Bronchitis	Nausea
Chest infection	Nose bleed
Conjunctivitis	Rash or rashes
Dermatitis	Runny nose
Diarrhea	Sinusitis
Difficulty sleeping	Stomach pain
Dizziness	Stevens Johnson syndrome
Ear infection	Symptoms of a cold

Ear problems	Tiredness
Erythema multiforme	Urticaria
Headache	Vomiting
Hepatitis	

Ingredients: Black iron oxide (E172), Croscarmellose Sodium, FD and C Blue 2 (indigo carmine, E132), Gelatin, Oseltamivir, Povidone, Pregelatinised maize starch, Red iron oxide (E172), Shellac, Sodium Stearyl Fumarate, Talc, Titanium dioxide (E171,) Yellow iron oxide (E172). The oral suspension has: Oseltamivir, Saccharin sodium (E954), Sodium benzoate (E211), Sodium dihydrogen citrate (E331 (a)), Sorbitol (E420), Titanium dioxide (E171), Tutti Frutti flavour, Maltodextrins (maize), Propylene glycol, Arabic gum (E414), Xantham gum (E415).

Various U.S. and U.N. agencies and media spread the word that the Avian Influenza, if it broke out, could have been as severe as the worldwide Spanish Influenza epidemic of 1918. They predicted hundreds of millions of deaths worldwide.

It did not happen.

Yet, they do it each and every year and it effectively sells a lot of vaccines of dubious efficacy, to an easily frightened public.

According to health consultant Jonathan Campbell, "The influenza, currently isolated in China, is a hemorrhagic illness. It kills many of its victims by rapidly depleting ascorbate (vitamin C) stores in the body, inducing scurvy and collapse of the arterial blood supply, causing internal hemorrhaging of the lungs and sinus cavities. Most people today have barely enough vitamin C in their bodies (typically 60 mg per day) to prevent scurvy under normal living conditions, and are not prepared for this kind of illness."[2]

*Some physicians would stand by and see their patient die
rather than use ascorbic acid (vitamin C) because
in their finite minds it exists only as a vitamin.*
Dr. Frederick R. Klenner

The International Medical Veritas Association's recommendation for flu prevention and treatment is simple and carries no side-effects: it consists of transdermal or topical application of magnesium chloride, zinc lozenges, vitamin C, and a properly hydrated body. Instead of weakening the body with the toxic substances found in Tamiflu, other drugs and vaccines, we strengthen the cells from their roots up. A great part of an individual's vulnerability to influenza is a combination of nutritional deficiencies and enormous toxic buildup from environmental poisons, toxic containing foods and noxious drugs.

The first physician to aggressively use vitamin C to cure diseases was Frederick R. Klenner, M.D.[3] beginning back in the early 1940's. Dr. Klenner consistently cured chicken pox, measles, mumps, tetanus and polio[4] with huge doses of the vitamin. Certainly if it is effective for these diseases it will help tremendously with the flu. From 1943 through 1947 Dr. Klenner reported successful treatment of forty-one more cases of viral pneumonia using massive doses of vitamin C.

Dr. Anthony S. Fauci, director of the National Institute of Allergy and Infectious Diseases, said, "You must prepare for the worst-case scenario. To do anything less would be irresponsible." Facuci is correct here, it pays to prepare but not with toxic vaccines or other drugs like Tamiflu. It makes sense medically to use the least toxic substances, for toxins only *increase* the chance we will get the flu. The flu-like side-effects of Tamiflu bears this line of reasoning out. The caution on pregnant women, or those considering becoming pregnant, should serve to turn away a good deal of the population thinking of using Tamiflu. If it is too toxic for mother and fetus, it is too toxic for you.

"Are we prepared today? Clearly no," said Dr. Hayden of the University of Virginia Health Science Center. From personal medicine cabinets to the White House, the focus recently is on a killer flu epidemic that is going to hit like a huge asteroid. If we trust our medical officials, we will prepare and if we do not trust these people, we still need to prepare because of the possibility that what they are warning us against is simply a long planned bio-chemical attack.[5,6]

Many believe that at heart the pharmaceutical companies have, for almost a century, used vaccines as a carrier of poisonous substances to earn profits over dead people's bodies. If we look at FluMist, another new nasal vaccine, we see a pharmaceutical company deliberately spilling into the environment contagious in-fluenza viruses.

The FluMist vaccine is a live virus vaccine that is given as a nasal spray. The virus can be shed from the nasal passages of a vac-cinated individual for up to 21 days after vaccination. The package insert for the FluMist vaccine clearly states, "Due to the possible transmission of vaccine virus, FluMist recipients should avoid be-ing in close (for example, within the same household) contact with immune-compromised individuals for three weeks following vac-cination."

Many might have missed seeing the congressional hearing on TV,[7] talking about bioterroism and hearing the blatant state-ment made that new vaccines, even if not yet proven safe, would be put out for use in the general population, as the threat of disease is seen to be greater than any risk that the use of these unproven vaccines might cause.

It should be noted that these policy decisions are political in nature, which are being made by politicians. The health and well-being of the public is *not* the main consideration.

These vaccines are being unleashed on us with the blessings and sanctioning of many government agencies. Docile, silent, and uninformed, we will be at great risk of harm. *The last thing the medical industrial complex is interested in is safe and natural methods of protection against viruses and the deficiencies that make us vulnerable.*

Magnesium depleted cells are cells in trouble. Introducing another toxic drug will not build their defenses or strength. *Removing* toxins will. Giving the body its vital nutrients will. The sad fact is that *most* of the world population is *needlessly* magnesium deficient.

The average severe flu lasts approximately 10 days, so we must question whether spending billions to stockpile Tamiflu to reduce that average by a day is really going to save anyone's life. The pharmaceutical industry, the CDC and the FDA of course see no problem with adding more toxic chemicals into the populations' blood streams.

Of course it pays to be prepared for the flu, so stock up your medicine cabinets with effective substances that will help, not hurt your family. It would also be worthwhile to detoxify our families during the warmer months so hidden toxicities like mercury, which is known to increase the dangers of influenza, would be less present in our children's bodies.

When a doctor's patient dies of influenza, it is not the virus that is deadly, but the patient's compromised immune system due to chronic, long-standing deficiencies of vitamin C and magnesium intake, combined with the drugs and vaccines that doctors use that send old and young alike to their graves. It's easy to blame doctors, but it is even easier to inform one's self, and make sure that our body has sufficient stores of the vital mineral nutrients that it needs, beginning with magnesium.

Why is the United States government so anxious to invest billions in Tamiflu and flu vaccines laced with mercury? *The Boston Globe* reported that alarmed infectious disease specialists have indicated that excessive use of Tamiflu and other antiviral drugs could lead to the emergence of flu strains that do not respond to anti-virals, making both avian and regular flu strains even more of a health threat. Yet another reason for using a natural protocol is raised and as usual, ignored by medical authorities.

Dr. Klenner is right though, medical authorities would stand-by and watch millions die rather than entertain the thought that it all could be stopped by using vitamin C and magnesium chloride. That would be just too simple. Don't sit idly by and let it happen to you, or someone that you love.

There should be no doubt that key government officials are in bed with the pharmaceutical companies. Defense Secretary Donald Rumsfeld, for example, seems likely to profit greatly from government purchase of Tamiflu, the drug developed by Gilead Sciences when Rumsfeld was president of the company.

He is reported to hold major portions of stock in Gilead.[8] Tamiflu was actually developed by Gilead, which then gave Roche Laboratories (www.roche.com) the exclusive rights to market and sell this drug.[9] Trust in our government's medical officials is in short supply.

References

1   Bird Flu infecting Vietnamese Girl found Resistant to Primary Drug

(See: http://www.indystar.com/apps/pbcs.dll/article?AID=/20051015/ NEWS01/510150466)

2   Campbell's recommendation: Begin increasing the amount of vitamin C that you take each day to very high levels, spread over the course of the day, in divided doses taken with meals. Start at 1000 mg per meal, and increase slowly to 2000-4000 mg per meal. (These are adult doses, modify by body weight for children.) Your optimal dose is just below the point where your body complains by giving you mild diarrhea. This is called the "bowel tolerance dose." Such doses are perfectly safe — vitamin C is natural to our bodies and needed for many body processes. Most people don't get nearly enough. Stock up on this vital nutrient - buy in powder form, 1-pound or 3-pound canisters (ascorbic acid form). Mix with water or fruit juice. Be sure to take vitamin C with food that will coat your stomach to prevent stomach upset, such as organic soymilk. (See: www.cqs.com/influenza.htm)

3   Dr. Klenner used massive doses of vitamin C for over forty years of family practice. He wrote dozens of medical papers on the subject. A complete list of them is in the *Clinical Guide to the Use of vitamin C*, edited by Lendon H. Smith, M.D., Life Sciences Press, Tacoma, WA (1988).

4   The Treatment of Poliomyelitis and Other Virus Diseases with vitamin C: Klenner, *Southern Medicine & Surgery*, July, 1949 "The treatment employed [in the poliomyelitis epidemic in North Carolina in 1948, 60 cases] was vitamin C in massive doses... given like any other antibiotic every two to four hours. The initial dose was 1000 to 2000 mg., depending on age. Children up to four years received the injections intramuscularly ... For patients treated in the home the dose schedule was 2000 mg. by needle every six hours, supplemented by 1000 to 2000 mg. every two hours by mouth ... dissolved in fruit juice. All patients were clinically well after 72 hours. Where spinal taps were performed, it was the rule to find a reversion of the fluid to normal after the second day of treatment.

5   A commentary in the *Journal of the Royal Society of Medicine* (Madjid et al. 2003) noted that influenza is readily transmissible by aerosol and that a small number of viruses can cause a full-blown infection. The authors continued: "the possibility for genetic engineering and aerosol transmission [of influenza] suggests an enormous potential for bioterrorism" The possible hostile abuse of influenza virus is seen as a very real threat by public health officials in the USA. $15 million was granted by the US National Institutes of Health to Stanford University to study how to guard against the flu virus "if it were to be unleashed as an agent of bioterrorism". Stanford University News Release 17 September 2003, (See: http://mednews.stanford.edu/ news_releases_html/2003/septrelease/bioterror%20flu.htm)

6   The resurrection of 1918 influenza has plunged the world closer to a flu pandemic and to a biodefense race scarcely separable from an offensive one, according to the Sunshine Project, a biological weapons watchdog. "There was no compelling reason to recreate 1918 flu and plenty of good reasons not to. Instead of a dead bug, now there are live 1918 flu types in several places, with more such strains sure to come in more places," says Sunshine Project Director Edward Hammond, "The US government has done a great misdeed by endorsing and encouraging the deliberate creation of

extremely dangerous new viruses. The 1918 experiments will be replicated and adapted, and the ability to perform them will proliferate, meaning that the possibility of man-made disaster, either accidental or deliberate, has risen for the entire world."

7   See: http://reform.house.gov/GovReform/Hearings/EventSingle.aspx?EventID=30083

8   See: www.gilead.com/wt/sec/pr_933190157/

9   See: http://sfgate.com/cgi-bin/article.cgi?file=/c/a/2005/06/24/MNGHTDE8LG1.DTL

17

||||||||||||||||||||||||||||||||||||||||||||||||||||||||||||||||||||

Inflammation and Systemic Stress

I
f we could ever understand the very basics of disease we could work efficiently toward furthering health. Public health officials, medical organizations and doctors moan and groan about epidemics of chronic diseases like diabetes, cancer, heart disease, autism and other neurological diseases, voicing their frustrations and their helplessness to do anything about it.

Inflammation and systemic stress are central attributes of many pathological conditions. Thus if we find a way to directly and safely reduce inflammation and systemic stress we have found a potent medical approach that would be effective across a wide range of pathologies.

Inflammation is the missing link to explain the role
of magnesium in many pathological conditions.

Dr. A. Mazur et al[1] have shown in experimentally induced magnesium deficiency in rats that after only a few days a clinical inflammatory syndrome develops and is characterized by leukocyte (white blood cell) and macrophage activation, release of inflammatory cytokines and excessive production of free radicals. "Magnesium deficiency induces a systemic stress response by activation of neuro endocrinological pathways," writes Dr. Mazur. "Magnesium deficiency contributes to an exaggerated response to immune stress and oxidative stress is the consequence of the inflammatory response," he continued.

Macrophages are a type of immune cell, found in the body's tissues and organs which rid the body of worn out cells and other debris, secrete powerful antigen-destroying chemicals, and play an important role in activating T-cells.

Cytokines are soluble molecules produced by cells that serve as mediators of intracellular reactions and biological response modifiers. Increases in extracellular magnesium concentration cause a decrease in the inflammatory response while reduction in the extracellular magnesium results in cell activation and hence, inflammation. Inflammation causes endothelial dysfunction and activated endothelium facilitates adhesion and migration of cancer cells.[2]

People with more magnesium and less copper
in their blood could reduce their risk of death
from cancer by as much as 50%,
says a new study from France.[3]

High serum levels of magnesium were also linked to a 40% lower risk of all-cause mortality, and a reduction of similar magnitude for cardiovascular deaths. The research is important because dietary surveys show that a large portion of adults do not meet the RDA for magnesium found naturally in green, leafy vegetables, meats, starches, grains and nuts, and milk.

"It's the first epidemiological study which shows synergistic effects on mortality of low serum zinc and high serum copper, as well as the effects of low serum zinc and low serum magnesium and these interactions may be physiopathologically plausible as these minerals are involved in the immune system, the inflammatory response and the oxidative damage," lead author Dr. Nathalie Leone from the Pasteur Institute, in Lille France.

After 18 years of follow-up, 339 subjects had died, with 176 due to cancer and a further 56 from cardiovascular disease (CVD). Serum analysis of magnesium, copper and zinc levels showed that the highest serum level of magnesium (0.85 millimoles per litre or more) compared to the lowest level (0.76 millimoles per litre or more), was associated with a reduction in the risk of all-cause mortality, death from cancer, and CVD of 40, 50 and 40%, respectively.

In the final analysis there is no single medicine or nutritional agent that has the power to both treat and prevent disease like magnesium. The proof is rock-solid and new studies are coming out every week to sustain this medical position.

It behooves us all to bring up our magnesium to optimal levels and keep them there and to pay attention to the same in our children. Magnesium acts as a general cell tonic while it reduces inflammation and systemic stress. Equally it is important in overall energy (ATP) production, hormonal and enzyme production and function, and just about everything else in biological life.

References

1   Mazur A., Maier J. A., Rock E., Gueux E., Nowacki W., Rayssiguier Y.
Magnesium and the inflammatory response: Potential physiopathological implications.
Arch Biochem Biophys. 2006 Apr 19; PMID: 16712775

2   Magnesium and inflammation: lessons from animal models] Clin Calcium.
2005 Feb;15(2):245-8. Review. Japanese. PMID: 15692164 [PubMed - indexed for
MEDLINE

3   The research, published in the journal *Epidemiology* (Vol. 17, pp. 308-314),
reports from the Paris Prospective Study 2, a cohort of over 4,000 men aged between
30 and 60 at the start of the study.

18

||||||||||||||||||||||||||||||||||||||||||||||||||||||||||||||||||||||||

Sexuality, Life, and Aging

Magnesium is important for us from before
we are born and until our last breath in life.

L ife, health, longevity and happiness are related con-
cepts and somehow our sexuality threads in the middle of that. Sex is part of our basic nature because our
bodies were conceived through the sexual union of a
man and woman, and on cellular levels, of male and
female cells.

Whenever sex is fulfilling, the chances of love and a joyful life
become greater. There is no denying that sex is one of the most plea-
surable activities available to human beings and that it is important
and central to life itself. It provides a universally recognized form
of enjoyment in all societies. However, for some very deep reasons
there also lies darkness across the world of sexuality.

Sex in particular has become a major source of anxiety and stress for many of us and this is not all our fault. Sometimes things beyond our control affect our sexuality, and thus our life itself is affected. Biologically all humans, male and female, have various levels of sex hormones. Males have more male hormones but still have some low or very low levels of female hormones. Females have more female hormones but still have some low or very low levels of male hormones. Testosterone is primarily a male sex hormone, and is thought to be the primary hormone responsible for the driving force of libido.

Emotional satisfaction during sex and coronary health are related. A healthy sexuality is potentially the best anti-stress medicine we have and when we embrace its full power and potentiality for healing we discover one of the greatest gifts God has given to us, an eternal fountain of youth and passion giver. With it we can bring a new life to earth and with it we can ever bring new life and love to our soul.

> *Sex can be both the fountain of youth*
> *and the fountain in which we replenish our love*
> *over and over again to the great joy of our beings.*

Many scientists believe that hormone-driven mechanisms in the central nervous system should dominate analysis of mammalian behaviors especially sexuality because sexuality is primarily hormone driven. Magnesium is necessary for normal sexual functioning, yet is glossed over in its importance in nervous and endocrine function necessary for good sexual performance.

In all likelihood pregnancy cannot be normal unless magnesium levels are adequate. The concentration of magnesium in the placental and fetal tissues increases during pregnancy. The requirements for this element in a pregnant woman's body generally exceed its supply; hence, pregnancy should be considered a condition of "physiological hypomagnesemia."[1]

At the time that life is waning, as we age
magnesium diminishes while calcium rises.

The fact that magnesium diminishes with age and illness should give us a clue to magnesium's vital place in life, health, sexuality and happiness. In this chapter we are going to attempt to unify the role magnesium has on basic human sexuality as well as aging and general health.

When we related magnesium in importance to the air we breathe perhaps we can begin to understand how important our daily magnesium input is. Five minutes without air spells doom. Deficient daily magnesium intake means serious trouble and speaks of disease when uncorrected. It means premature aging, declining fertility and sexual potency and ability to conceive new life.

Magnesium is found in higher levels in semen than
serum. Orhan Deger from the Ataturk University
in Turkey found infertile men had about half the
semen magnesium as fertile men.[2]

The role of magnesium in our lives continues its importance when we are in the womb. When pregnant, magnesium helps build and repair body tissue in both mother and fetus. A severe deficiency during pregnancy may lead to pre-eclampsia,[3] birth defects and infant mortality. Magnesium relaxes muscles and research suggests that proper levels of magnesium during pregnancy can help keep the uterus from contracting until week thirty-five. Dropping magnesium levels at this point may start labor contractions. This has been treated by allopathic medicine with magnesium sulfate but this has led to problems we discussed in the chapter that compares magnesium sulfate to magnesium chloride.

In animal studies it has been shown that magnesium plays a role in ovule maturation, sperm viability and fecundation. In rats, pregnancy cannot be normal unless the food contains an adequate supply of magnesium. Severe or mild deficiencies affect the site of fetal implantation and, if they are prolonged, lead to abortion in the first instance and pathological disorders in the latter.[4]

Some 43% of women suffer with sexual dysfunction, compared to 31% of men, according to University of Chicago researcher Dr. Edward Laumann.[5]

Some $2 to $3 billion will be spent within the next ten years on products aimed at improving the sex lives of these women. Female sexual dysfunction is characterized by a lack of desire, arousal and orgasm. Lack of desire is the chief complaint among women, affecting about one-third of them at some point in their lives, says Cindy Meston, assistant professor of clinical psychology at the University of Texas at Austin.[6] How much of this dysfunction is related to magnesium deficiency is anyone's guess but when upward of 70% of all people are deficient in this precious metal of life it is not too hard to make the connection.

Magnesium and DHEA

*DHEA -S levels were significantly lower
in the men with sexual dysfunction.*[7]

For normal sexual function, we need both healthy organs and a balanced, working endocrine system, producing the necessary hormones. Cholesterol cannot be synthesized without magnesium and cholesterol is a vital component of many hormones.

Aldosterone is one such hormone, and helps to control the balance of magnesium and other minerals in the body. Interestingly aldosterone needs magnesium to be produced and it also regulates magnesium's balance.[8] Cholesterol, that most maligned compound, is actually crucial for health and is the mother of hormones from the adrenal cortex, including cortisone, hydrocortisone, aldosterone, and DHEA.

*DHEA-S is a natural substance that has been shown
to improve sexual desire and overall quality of life.*

Flaunted as a "fountain of youth," DHEA is a natural hormone produced by the adrenal glands. This hormone rises at puberty, is at its apex till the age of thirty and drops down drastically afterwards. At the last stage of life i.e. between 70 and 80 the quantity of the hormone produced is only 5% of the amount churned out at the age of 30. Another factor that contributes significantly in DHEA loss is chronic stress. The hormone helps aging men and women increase muscle strength and lean body mass, regain youthful vitality, and just feel better.

Also known as "mother of all steroid hormones" DHEA is converted in the body into several different hormones, including estrogen and testosterone. But according to some medical scientists replacement of diminishing DHEA with synthetic DHEA can lead to health hazards including cancer, heart disease, diabetes, Alzheimer's, Parkinson's, and lupus.[9]

Much of the popular and scientific interest in DHEA stems from our culture's emphasis on youth. If levels of this hormone decline with age, the thinking goes, we could avoid the health problems that accompany aging — or even extend our lifespan — by keeping DHEA levels high.[10] Many people are already taking DHEA just in case this turns out to be true. That wouldn't be a problem if this substance were as safe as vitamin C. But as a potent steroid hormone, DHEA has the potential for far-reaching side effects throughout the body.

Dehydroepiandrosterone (DHEA) is a steroid hormone produced by the adrenal gland and ovaries and converted to testosterone and estrogen. DHEA can be purchased over the counter in supplement form. In one small study published in the *New England Journal of Medicine* (Sept. 30, 1999), women who took 50 mg of DHEA daily noticed a significant increase in sexual interest. However, most DHEA products lining the store shelves recommend taking only 25 mg per day. Because of its potential for heart attacks and breast cancer and masculating side-effects such as facial hair, synthetic DHEA is dangerous thus it would serve us if a natural way of increasing DHEA levels were to be found.

Magnesium chloride, especially when applied transdermally, is reported by Dr. Norman Shealy to gradually increase DHEA in a natural way.[11] Dr. Shealy has determined that when the body is presented with adequate levels of magnesium at the cellular level, the body will begin to naturally produce DHEA, as well as DHEA-S.

Because of the problems with supplementing directly with DHEA, transdermal magnesium chloride mineral therapy holds great promise for DHEA hormonal supplementation because taking magnesium could counteract the heart attack and stroke hazards of synthetic DHEA hormone replacement therapy.

The DHEA Adverse Effects (Via Synthetic Supplementation)

• In women, excess DHEA can produce unwanted side effects like flare-ups of acne.

• Higher levels of DHEAS appear to be linked to an increased risk for ovarian cancer (in both pre- and postmenopausal women) as well as breast cancer (in postmenopausal women).

• In addition, because DHEA is converted to testosterone, there is concern that long-term use of DHEA supplements may raise the risk for prostate cancer or, in those who have the disease, hasten its progression.

It appeared that for old men with a deficit of testosterone, Mg(2+) supplementation during treatment with DHEA can increased the free testosterone concentration and its biological effect.[12]

In men, too little testosterone may cause difficulty obtaining or maintaining erections, but it is not clear whether testosterone deficiencies interfere with female sexual functioning apart from reducing desire. Magnesium's ability to increase DHEA levels, thus testosterone levels in both men and women, demonstrates the pathway between transdermal magnesium chloride and sexual function and libido.

Testosterone plays a vital role in many functions throughout the body:

- Muscle mass maintenance
- Blood sugar regulation
- Bone density maintenance
- Oxygen uptake by cells throughout the body
- Immune surveillance
- Healthy red blood cell production
- Cholesterol regulation
- Heart muscle maintenance
- Sexual desire and performance
- Neurological activity

Women need and respond to much lower levels of testosterone increases than men do with significant increases in libido.

For the first few years of life, the adrenals make very little DHEA. Around age six or seven, they begin churning it out. Production peaks in the mid-20s, when DHEA is the most abundant hormone in circulation. At all ages, men tend to have higher DHEA levels than women.

Serum magnesium falls with the cyclic increase in estrogen secretion.[13]

After being secreted by the adrenal glands, it circulates in the bloodstream as DHEA-sulfate (DHEA-S) and is converted as needed into other hormones. DHEA also modulates immunity. A group of elderly men with low DHEA levels who were given 50 mg of DHEA per day for 20 weeks experienced a significant activation of immune function. Postmenopausal women have also shown increased immune functioning in just three weeks when given DHEA in double-blind research.

DHEA appears to protect every part
of the body against the ravages of aging.

Published studies link low levels of DHEA to aging and diseased states. Specifically, a deficiency of DHEA has been found to correlate with:

- Chronic inflammation

- Immune dysfunction

- Depression

- Rheumatoid arthritis

- Type-II diabetic complications

- Greater risk for certain cancers

- Excess body fat

- Cognitive decline

- Heart disease in men

- Osteoporosis

In 1981, the Life Extension Foundation introduced DHEA in an article that described the multiple anti-aging effects this hormone might produce. DHEA did not establish scientific credibility until 1996, when the New York Academy of Sciences published a book titled *DHEA and Aging* and summarized the concept of DHEA replacement in their journal *Aging*.

The role that magnesium plays in the transmission of hormones (such as insulin, thyroid, estrogen, testosterone, DHEA, etc.), neurotransmitters (such as dopamine, catecholamines, serotonin, GABA, etc.), and mineral electrolytes is a strong one. Research concludes that it is **magnesium status that controls cell membrane potential** and through this means controls uptake and release of many hormones, nutrients and neurotransmitters.

It has been found that DHEA levels may be raised in human beings through transdermal absorption of magnesium without the application of exogenous supplements of DHEA, DHEA-S, or their corresponding salts. Treatments consisting of applying a composition comprising a therapeutically effective amount of magnesium chloride.[14]

Since DHEA is one of the primary bio-markers for aging, the long range effect of large doses of magnesium in a usable form is to significantly raise DHEA levels and thus produce true age reversal results. But if you are an obese postmenopausal woman or a male with prostate cancer, then you certainly would not want to supplement DHEA directly. DHEA as the master hormone, (actually it can be considered more as a prohormone, a precursor, which is true for magnesium as well), when produced at sufficient levels, will induce the production of other hormones whose depletion can be associated with many symptoms of aging.

Through the use of magnesium oil, women have reported complete abatement of menopausal symptoms and some have even returned to their menstrual cycle and the possible explanation for this is that rising magnesium levels brings up DHEA levels.

Local Effects

Magnesium is essential for life as it participates in over 325 enzyme reactions and thus when we treat the deficiency we can expect more energy, strength and stamina and even increases in sexual energy. Magnesium chloride often produces systemic excitation. It also has a strong local effect, which is easily experienced when applied to sore muscles. Even more pronounced is the effect when we apply magnesium chloride to the mucus membranes.

"Spray the magnesium oil into and around the vaginal area, give it a try, see what happens. Amazing for me when I did this by accident, within an hour or so and now within minutes, I can hardly stand not to have sex."

Dr. Robert Ornstein and Dr. David Sobel, regional director for preventive medicine for Kaiser Permanente, the world's largest health maintenance organization, say that the physiology of sexual excitement in both men and women depends on the ability of blood to flow into the genital glands and into the supporting tissues and muscles of the pelvis. They also say that "it is the full engorgement of blood in the muscles and tissues of the pelvis that eventually triggers orgasm. One reason that women in general may be slower to orgasm than men is because they have a far more complex system of arteries, veins, and capillaries in the pelvic area than men do that must be fully engorged for orgasm to occur." And in fact pharmaceutical companies, in searching for a female equivalent of Viagra, are focused on developing a drug that increases blood flow to the female genitals, resulting in vaginal lubrication and relaxing vaginal muscles.

"A person who feels uncomfortable with his or her sexuality will unconsciously tense the muscles in the genitals, thighs, anus, and buttocks, preventing the free flow of blood into the area and thereby limiting the possibility for complete arousal. But a person who can relax and allow energy in the form of blood to flow into the pelvic area and genitals will enjoy the sensations of building excitement," wrote Drs. Ornstein and Sobel.

In men, decreased levels of magnesium gives rise to vaso-constriction from increased thromboxane level, increased endo-thelial intracellular calcium, and decreased nitric oxide. This may lead to premature emission and ejaculation processes. Magnesium is also probably involved in semen transport.[15]

Magnesium chloride delivers a relaxing and opening effect to general body physiology. Magnesium acts peripherally to produce vasodilation. With low doses only flushing and sweating occur, but larger doses actually will cause a lowering of blood pressure. It is known that magnesium causes nitric oxide independent coronary artery vasodilation in humans.[16]

Magnesium is thought to behave like a calcium channel blocker at the cellular level thus helps to maintain the elasticity of our arteries. Magnesium prevents the deposition of calcium along the arterial wall at points of micro-injury, which is the crucial role it plays in the prevention of both atherosclerosis and arteriosclerosis. Magnesium ions appear to reduce vascular resistance. Magnesium also is important in the maintenance of healthy muscles. The heart muscle itself benefits from an adequate supply of available magne-sium. For these reasons, magnesium is critical to the maintenance of a healthy heart. We would expect the same in all the muscles including those in the pelvic and genital areas.

We should not be surprised that magnesium chloride, a strong cell tonic, one that releases energy production by boosting ATP levels as well as hundreds of enzyme reactions, would have this effect.

Low pituitary function may lead to decreased development of the sexual organs, early menopause in women, and impotence in men. Weak adrenals may reduce the desire and strength for sex and increase sensitivity to stress. Low thyroid may cause a lack of desire or capacity for sex. In men, low testicular function decreases sex drive and sperm production. In women, low estrogen slows sexual maturity, decreases breast size, and retards egg maturation. Estrogen-progesterone imbalance can create many menstrual cycle variations and symptoms. According to Dr. Mildred Seelig, "The high magnesium content in the spinal fluid is that the mineral is necessary for balancing the stimulant effect of body hormones. The purpose of thyroid, gonadal, adrenal and other hormones is to charge up or excite the body. Magnesium and some other substances tend to slow down and relax the system, thus regulating the hormones and achieving a happy medium."[17]

Thyroid hormone is intimately associated with regulation of energy production and mitochondrial function. Indeed, mitochondria are equipped with thyroid hormone receptors. According to Biotics Research Corporation enzyme utilization of ATP generated by mitochondria requires complex formation with magnesium, generally in a 1:1 ratio. Magnesium is essential for protein synthesis, cell replication and activation of the sodium-potassium pump, as well as regulation of calcitonin and parathyroid hormone.[18]

Conclusion

In Taoist health & longevity systems, there has always been a strong link between sexual potency, immune response, cerebral function, and longevity. Magnesium is the most essential nutrient mineral to all of these factors. We will live healthier, happier and longer lives if we make sure we completely satisfy our body's need for magnesium. Not only do our cells need an abundant reserve of magnesium in the body, we need to absorb a sufficient replenishing amount each and every day.

References

1   Semczuk M, Semczuk-Sikora A. New data on toxic metal intoxication (Cd, Pb, and Hg in particular) and Mg status during pregnancy. *Med Sci Monit*. 2001 Mar-Apr;7(2):332-40.

2   *Magnesium*, 1988, vol 7.

3   Preeclampsia, also known as toxemia, is a complex disorder that affects about 5 to 8% of pregnant women. You're diagnosed with preeclampsia if you have high blood pressure and protein in your urine after 20 weeks of pregnancy. The condition most commonly shows up after you've reached 37 weeks, but it can develop any time in the second half of pregnancy, as well as during labor or even after delivery (usually in the first 24 to 48 hours). Preeclampsia causes your blood vessels to constrict, resulting in high blood pressure and a decrease in blood flow that can affect many organs in your body, such as your liver, kidneys, and brain. When less blood flows to your uterus, it can mean problems for your baby, such as poor growth, decreased amniotic fluid, and placental abruption — when the placenta separates from the uterine wall before delivery. In addition, your baby may suffer the effects of prematurity if you need to deliver early to protect your health.

4   Stolkowski J. Magnesium in animal and human reproduction *Rev Can Biol*. 1977 Jun; 36 (2):135-77.

5   Study by Edward O. Laumann, PhD; Anthony Paik, MA; Raymond C. Rosen, PhD; *JAMA* 2/9/99.

6   Christopher Gearon; *The Search For A Female Viagra*; http://health.discovery. com/centers/womens/viagra/viagra.htm

7   Basar MM. et al. *Relationship between serum sex steroids and Aging Male Symptoms score and International Index of Erectile Function*. Department of Urology, University of Kirikkale, Kirikkale, Turkey. Urology. 2005 Sep;66(3):597-601.

8   A deficiency in magnesium causes hyperplasia of the adrenal cortex, elevated aldosterone levels, and increased extracellular fluid volume. Aldosterone increases the urinary excretion of magnesium; hence, a positive feedback mechanism results, which is aggravated since there is no renal mechanism for conserving magnesium.

9   See: www.prescriptionsfirst.com/health_articles/dhea_trying_to_make_evergreen_lives.html

10   Some reports have suggested that DHEA might reduce the risk of heart disease perhaps by lowering cholesterol levels. Systemic lupus erythematosus (SLE) an autoimmune disease has been linked to abnormalities in sex hormone metabolism. Supplementation with very large amounts of DHEA (200 mg per day) improved clinical status and reduced the number of exacerbations of SLE in a double-blind trial A preliminary trial has confirmed the benefit of 50-200 mg per day of DHEA for people with SLE. People infected with HIV and those with insulin-dependent diabetes, congestive heart failure, multiple sclerosis, asthma, chronic fatigue syndrome, rheumatoid arthritis, osteoporosis, and a host of other conditions have been reported to have low levels of DHEA. Most, studies have found that people with Alzheimer's have lower blood DHEA-S levels than do people without the condition. Thus all of these conditions will respond to transdermal magnesium chloride therapy, which clinically would be a safer way to raise DHEA levels.

11   See: www.betterway2health.com/cwr-dhea.htm

12   Andre, C. et al. Testimony of the correlation between DHEA and bioavailable testosterone using a biochromatographic concept: effect of two salts. *J Pharm Biomed Anal.* 2003 Dec 4;33(5):911-21.

13   Dahl, 1950; Nida and Broja, 1957; Goldsmith, 1963; Goldsmith et al., 1970; Goldsmith, 1971.

14   United States Patent Application 20050089581 Shealy, C. Norman April 28, 2005 Magnesium containing compositions and methods for enhancing dehydroepiandrosterone levels.

15   Omu AE, Al-Bader AA, Dashti H, Oriowo MA. Magnesium in human semen: possible role in premature ejaculation. Department of Obstetrics and Gynaecology, Faculty of Medicine, Kuwait University, *Safat. Arch Androl.* 2001 Jan-Feb;46(1):59-66.

16   H Teragawa, M Kato, T Yamagata, H Matsuura, G Kajiyama. The First Department of Internal Medicine, Hiroshima University School of Medicine, 1-2-3 Kasumi, Minamiku, Hiroshima, Japan

17   Magnesium, The Nutrient That Could Change Your Life: http://www.mgwater.com/rod07.shtml

18   See: www.bioticsresearch.com/PDF/Thyrostim.pdf

19

||||||||||||||||||||||||||||||||||||||||||||||||||||||||||||||||||||||

Transdermal Magnesium Mineral Therapy in Sports Medicine

*Magnesium nutrition is an area that
no serious athlete can afford to overlook.*

espite magnesium's pivotal role in energy production, many coaches and athletes remain unaware of its critical importance in maintaining health and performance. Research suggests that even small shortfalls in magnesium intake can seriously impair athletic performance.[1]

Magnesium deficiency reduces metabolic efficiency, increases oxygen consumption and heart rate required to perform work, all things that would take the edge off of athletic performance (not to mention carrying out the functions of normal life). The last thing any trainer or sports doctor wants to see is their athletes lose their competitive edge. Not performing to full capacity because of the lack of a mineral like magnesium is simply not an option for winners. Athletic endurance and strength performance increases significantly when a large amount of magnesium is supplemented transdermally/topically and orally.

A magnesium shortfall can cause a partial uncoupling of the respiratory chain, increasing the amount of oxygen required to maintain ATP production.

Athletes, who might be expected to take greater care with their diets, are not immune from magnesium deficiency. For example, studies carried out in 1986/87 revealed that gymnasts, football and basketball players were consuming only around 70% of the RDA,[2] while the intake of female track and field athletes was even lower, as low as 59% of the RDA.[3]

Evidence shows that a magnesium shortfall boosts the energy cost, and hence oxygen use, of exercise during activities like running or cycling.[4] One study of male athletes supplemented with 390 mg of magnesium per day for 25 days resulted in an increased peak oxygen uptake and total work output during work capacity tests.[5]

Sub-optimum dietary magnesium intakes impairs athletic performance.

The regular advice given athletes is to make a conscious effort to increase the proportion of magnesium-rich foods in his or her diet. Even a simple change like eating more whole grain products and boosting your intake of vegetables, nuts and seeds can make an impact. But that is not enough, not for an athlete who loses magnesium much faster than the average person. It is not even enough today for a regular person.

If an athlete is not eating a heavy diet of pumpkin seeds (roasted), spirulina, almonds, brazil nuts, sesame seeds, peanuts, walnuts or rice (whole grain brown), the only common foods with over 100 milligrams of magnesium content for every 100 grams, it is not really in the realm of possibility that sufficient magnesium would be consumed, and this is assuming that the soil these foods are grown in has sufficient levels of the mineral. Add white bread and other forms of processed, or junk food—even endorsed by some of today's top athletes—and you can see why athletes are unlikely to consume enough magnesium through dietary sources alone.

It is commonly thought that magnesium intakes above the RDA are unlikely to boost performance, but there is no evidence to support the assertion. First, RDAs are almost universally *understated*, even for the general population. In other words, they represent minimums that should be taken for the maintenance of health. For athletes RDA's are sure guides to failure for they do not take into account the extra demands and needs of an athlete's body. When it comes to magnesium an athlete should be thinking many times the RDA if he or she wishes to maximize athletic performance. Having an adequate store of this vital, natural mineral will ensure total system availability, without the downside of synthetic agents, such as steroids, which create physiological imbalances.

Studies have shown that dietary supplementation with 30 mg of zinc and 450 mg of magnesium daily can elevate testosterone levels up to 30%.

Dr. Lorrie Brilla, at Western Washington University, recently reported that magnesium and zinc, when supplemented orally, significantly increase free testosterone levels and muscle strength in NCAA football players.[6] In another study, young athletes supplemented with 8 mg of magnesium per kilo of body weight per day, experienced significant increases in endurance performance and decreased oxygen consumption during standardized, sub-maximal exercise.[7]

Dr. Brilla reported that during an eight-week spring training program athletes had 2.5 times greater muscle strength gains than a placebo group.[8] Any athlete looking to gain strength, increase athletic performance, and muscle mass should consider greatly increasing their magnesium intake, as well as zinc.

> *Muscle endurance and total work capacity,*
> *declines rapidly with nutritional deficiency in the*
> *area of key minerals like zinc and magnesium.*

"Magnesium is essential to a diet for people who are under a lot of stress or want to experience the ultimate rush," says Dr. James Thor, National Director of Extreme Sports Medicine. "Several reasons, one is if you are working out in a gym, or continual stress excessive amounts of lactic acid in the muscle have been linked to higher levels of anxiety," Dr. Thor adds. Large amounts of magnesium are lost when a person is under stress.

The combination of heat and magnesium chloride increases circulation and waste removal and this principle can be applied during breaks in competition as well as after the game in deeply relaxing baths similar to Epsom salt baths, but much stronger. A magnesium chloride bath helps draw inflammation out of the muscles and joints. Dr. Mark Steckel recommends a hot bath with Epsom salts (magnesium sulfate) after a long run when the muscles are just aching. He also recommends soaking once a week "as a treat to your legs, just to keep them happy!" Switching to magnesium chloride takes the experience to an entirely new level of therapeutics.

Transdermal magnesium chloride mineral therapy enhances recovery from athletic activity or injuries.

A whole new world of sports medicine is going to explode onto the scene when athletes and coaches find out that magnesium chloride from *natural* sources is available for topical use. In this new and exciting breakthrough in sports medicine coaches can now treat injuries, prevent them, and increase athletic performance all at the same time.

Magnesium chloride, when applied directly to the skin is transdermally absorbed. Transdermal magnesium chloride mineral therapy is ideal for athletes who need high levels of magnesium. Oral magnesium is much less effective than transdermal magnesium in the treatment of injuries and tired worn out muscles.

Until now it was thought that the best forms of supplemental magnesium were the ones chelated to an amino acid (magnesium glycinate, magnesium taurate) or a krebs cycle intermediate (magnesium malate, magnesium citrate, magnesium fumarate). These forms seem to be better utilized, absorbed, and assimilated. Some have correctly advised to stay away from oral intake of inorganic forms of magnesium like magnesium chloride (taken orally) or magnesium carbonate because they may not be absorbed as well and may cause gastric disturbances. But now we have a magnesium chloride lotion/bath salt that can be applied directly to the skin so dosage levels can be brought up safely to high levels without diarrhea and problems with absorption.

Maximal contraction of the quadriceps is
positively correlated to serum magnesium status.[9]

Dr. Jeff Schutt insists that hamstring injuries can at least partially be avoided through nutritional support because contraction and relaxation is dependent on adequate cellular levels of magnesium. "A shortened hamstring is a result of lack of available magnesium," he says. Liquid magnesium chloride can be simply sprayed and rubbed into a sore Achilles tendon to decrease swelling. And soaking the feet in a magnesium chloride foot bath is the single best thing — apart from stretching — that you can do for yourself to protect from, or recover from hamstring and other injuries. The only thing better is a full body bath or to have a massage therapist use it to rub it in as they work deeply on the muscles.

The heavy use of magnesium for athletic performance
will be enough to make a difference between
winning and losing on a regular basis.

Magnesium is the single most important mineral to sports nutrition. Adequate magnesium level will help your body against fatigue, heat exhaustion, blood sugar control, and metabolism. It also offers part of the secret why athletes die young — magnesium levels in tissue analysis are usually very low, and often mercury very high in athletes who have heart attacks. Congestive heart failure patients have recently been reported to have 22,000 times more mercury and 14,000 times more antimony in their hearts.[10]

Zinc, chromium and selenium in addition to magnesium are excreted in the sweat[11,12] or in the actual accelerated metabolism of strenuous exercise and are difficult to replenish.[13] When we sweat, we lose more than just water. Other components of sweat include electrolytes, principally sodium and magnesium. Loss of magnesium by sweating takes place at an accelerated pace when there is a failure in sweat homeostasis, a situation which arises when exercise is made in conditions of damp atmosphere and high temperature.[14]

Increased energy expenditure causes an increase in magnesium requirements. Selenium is important in that it neutralizes the toxic effects of mercury. This is especially important for athletes who have a mouth full of mercury containing dental amalgam.[15] Beware the sports people who say that the amount of magnesium lost through sweat is negligible, making magnesium supplementation unnecessary.[16] Dr. Sarah Mayhill says, "Heavy exercise also makes you lose magnesium in the urine and explain why long distance runners may suddenly drop dead with heart arrhythmias." Magnesium intake is most often marginal at best and heavy exercise is a factor that is particularly likely to expose athletes to magnesium deficit through metabolic depletion linked to exercise.

Magnesium depletion and deficiency play a role
in the pathophysiology of physical exercise.[17]

Many in sports medicine think that supplements should only be taken when there is proof that the diet cannot provide the quantities of nutrients needed and that supplements require a proper medical diagnosis and should only be prescribed by the sports physician and dietician in writing. Some go as far as insisting that fitness coaches and conditioning staff should not prescribe any supplements. But trainers need to be aware of anything that would enhance or help reduce the amount of time for rehabilitation due to an injury. The job of trainers and coaches is to prevent injuries or to get the players well as quickly as possible.

Everyone involved in athletics need to be acutely aware that the medical industrial complex is not going to act in the best interests of athletes. Many are becoming more conscious about how good ideas and sound natural medicine are being professionally suppressed by intricate campaigns of discreditation, spun by the vested interests of corporate science and backed by the pharmaceutical industry and even the government which is in bed with the drug companies.

The last thing they want athletes or the general public to know is that it is now virtually impossible to receive needed and necessary nutrition from foods grown from modern agricultural methods. Nutritional values of foods have been dropping precipitously over the last fifty years and the increasing toxic exposures put additional demands on an athlete's nutritional status. This is especially true with magnesium.

There is virtually no one that cannot benefit greatly from increasing daily magnesium intake. In terms of health and longevity magnesium is essential. For the professional athlete it means the difference between winning and losing, and in some cases, living and dying.

References

1   In a very tightly controlled three-month US study the effects of magnesium depletion on exercise performance in 10 women were observed – and the results make fascinating reading . In the first month, the women received a magnesium-deficient diet (112mgs per day), which was supplemented with 200mgs per day of magnesium to bring the total magnesium content up to the RDA of 310mgs per day. In the second month, the supplement was withdrawn to make the diet magnesium-deficient, but in the third month it was reintroduced to replenish magnesium levels. At the end of each month, the women were asked to cycle at increasing intensities until they reached 80% of their maximum heart rate, at which time a large number of measurements were taken, including blood tests, ECG and respiratory gas analysis. The researchers found that, for a given workload, peak oxygen uptake, total and cumulative net oxygen utilization and heart rate all increased significantly during the period of magnesium restriction, with the amount of the increase directly related to the extent of magnesium depletion. In plain English, a magnesium deficiency reduced metabolic efficiency, increasing the oxygen consumption and heart rate required to perform work – exactly what an athlete doesn't want!

2   *J Am Diet Assoc*;86: 251–3 (1986) and *Nutr Res*;7:27–34 (1987)

3   *Med Sci Sports Exerc*; 18(suppl):S55–6 (1986)

4   *J Appl Physiol* 65:1500-1505 (1988)

5   *Endocrinol Metab Clin N Am* 22:377-395 (1993)

6   Brilla, Lorrie. ACSM journal, *Medicine and Science in Sports and Exercise*, Vol. 31, No. 5, May 1999.

7   *Med Exerc Nutr Health* 4:230-233 (1995)

8   Pre and post leg strength measurements were made using a Biodex isokinetic dynamometer." The strength of the ZMA group increased by 11.6% compared to only a 4.6% increase in the placebo group.

9   G. Stendig-Lindberg, et al., "Predictors of maximum voluntary contraction force of quadriceps femoris muscle in man. Ridge regression analysis," *Magnesium 2* (1983): 93-104.

10   Frustaci, A., et al. Marked Elevation of Myocardial Trace Elements in Idiopathic Dilated Cardiomyopathy Compared With Secondary Dysfunction. Department of Cardiology, Catholic University, Rome Italy *Journal of the American College of Cardiology*. Vol. 33, No. 6, 1999, pp. 1578-1583: A large increase (>10,000 times for mercury and antimony) of TE concentration has been observed in myocardial but not in muscular samples in all patients with IDCM. Patients with secondary cardiac dysfunction had mild increase (<5 times) of myocardial TE and normal muscular TE. In particular, in patients with IDCM mean mercury concentration was 22,000 times (178,400 ng/g vs. 8 ng/g), antimony 12,000 times (19,260 ng/g vs. 1.5 ng/g), gold 11 times (26 ng/g vs. 2.3 ng/g), chromium 13 times (2,300 ng/g vs. 177 ng/g) and cobalt 4 times (86.5 ng/g vs. 20 ng/g) higher than in control subjects.

11   C. Consolazio, et al., "Excretion of sodium, potassium, magnesium, and iron in human sweat and the relation of each to balance and requirements," *J. Nutr* 79 (1963): 407-415.

12   R. McDonald and C. Keen, "Iron, zinc, and magnesium nutrition and athletic performance," *Sports Med.* 5 (1988): 171-184.

13   P. Deuster, et al., "Magnesium homeostasis during high-intensity anaerobic exercise in men," *J. Appl. Physiol.* 62 (1987): 545-550.

14   According to Dr. Jeffrey Sankoff, "Because our bodies can only function within a narrow range of temperature, mechanisms exist for cooling. The most important of these mechanisms is the production of sweat. When sweat is formed on the skin, the heat from the body evaporates the water and energy is dissipated. However, if it is very hot sweating becomes less efficient as the air — rather than heat generated by the body — evaporates the sweat. And in humid conditions water evaporation slows, so sweating becomes less effective." See: www.insidetri.com/train/tips/articles/2218.0.html

15   The average size amalgam filling contains approximately 750,000 micrograms of mercury (Hg) which releases part of that everyday for as long as the filling is in a person's mouth. A microgram (mcg) is 1/1,000 of a milligram in weight or one-millionth of a gram. A milligram (mg) is 1/1,000 of a gram by weight. People with amalgam are exposed to from tens to several hundreds of micrograms of mercury per day depending on how many fillings are in their mouth, how old the fillings are, how much a person brushes their teeth, chews and eats, the bacteria count in the mouth, and even the temperature of the body. Dr. Murry Vimy, professor of dentistry says, "It is estimated that the average individual, with eight biting surface mercury fillings, is exposed to a daily dose uptake of about 10 micrograms mercury from their fillings. According to Dr. Magnus Nylander, "Data suggest that approximately 19 to 20% of the general population may experience sub-clinical CNS and/or kidney function impairment as a result of the presence of amalgam fillings." Dr. Robert Gammal states, "Mercury from amalgam fillings has been shown to be neurotoxic, embryotoxic, mutagenic, teratogenic, immunotoxic and clastogenic. It is capable of causing immune dysfunction and auto-immune diseases." It is important to remember that mercury toxicity is a retention toxicity that builds up during years of exposure. The toxicity of a singular level of mercury is greatly increased by current or subsequent, low exposures to lead or other toxic heavy metals.

16   Y. Rayssiguier1, C. Y. Guezennec, and J. Durlach. INRA, Laboratoire des Maladies Métaboliques, France: Urinary Mg losses during an endurance event could play a role in this depletion but are often reduced, reflecting renal compensation. Loss of Mg by sweating takes place only when there is a failure in sweat homeostasis, a situation which arises when exercise is made in conditions of damp atmosphere and high temperature. Stress caused by physical exercise is capable of inducing Mg deficit by various mechanisms. A possible explanation for decreased plasma Mg concentration during long endurance events is the effect of lipolysis. Since fatty acids are mobilized for muscle energy, lipolysis would cause a decrease in plasma Mg.

17   Y. Rayssiguier1, C. Y. Guezennec, and J. Durlach. INRA, Laboratoire des Maladies Métaboliques, France

20

||||||||||||||||||||||||||||||||||||||||||||||||||||||||||||||||||||||||||

Menopause and Premenstrual Syndrome

Endocrine Function and Magnesium

Every day the same type of conversation takes place between women going through menopause and their doctors. Afterwards doctors usually write out prescriptions for estrogen pills or patches, saying they will replace the hormones that a woman's body ought to be making. The doctor promises these medicines will cure her hot flashes, slow her bone loss, and reduce her risk of a heart attack. Unfortunately we find out:

> *The risk of having a blood clot was close to 30%*
> *more for a woman on estrogen vs. not on estrogen.*[1]
>
> Dr. J. David Curb

Estrogen therapy appears to increase the risk of blood clots in the veins of postmenopausal women who have had their uterus removed. These latest results from the Women's Health Initiative (WHI) were unexpected, even to the study's lead author. "It surprised us all how few benefits have come out of this and how many negatives," said Dr. J. David Curb, a professor of geriatric medicine at the University of Hawaii. The study appeared in the April 10, 2006 issue of the *Archives of Internal Medicine*.

These same women ask if the pills cause cancer. The doctor acknowledges that there is an increased risk of uterine and breast cancer, but argues that the benefits to the heart and bones are worth taking the chance. Of course there is concern about other risks from these medications like strokes and water retention, among others.

*Like animals lured into a snare by a trail of crumbs,
women have been cajoled with scientific studies,
media advertising, patient hand books and
drug samples to accept Hormone Replacement
Therapy as a magic potion.*

Sherill Sellman

Hormone Replacement Therapy (HRT) does not do justice to the finely tuned hormone system[2] that operates throughout a woman's life. In reality, hormone levels may begin to change in the 30s, as a woman enters a period called perimenopause. In the decades leading up to menopause, small hormonal imbalances can exist, so by the time menopause sets in, a woman may have already experienced close to 20 years of hormonal imbalance.

Hormone replacement actually can cause the body to slow down the production of its own natural hormones, including melatonin, DHEA, progesterone and human growth hormone (HGH). HRT does not treat the cause of any problem, it only addresses

— and perpetuates — the symptoms. Adding hormones artificially is a form of medical roulette because you never really know how the finely balanced hormonal system will respond.

> My MD put me on hormone therapy, a combination of estrogen and progesterone and the effects on me were profound. Within the first month .my feet and legs swelled up all the way to my knees. I could not get shoes on, (but the hot flashes were gone) and when I stopped the hormones, it took almost a full month for the swelling to dissipate and my feet and legs to get back to normal. I experienced breakthrough bleeding, which was told to me might occur, but everyday?
>
> Claudia French RN, LPHA

In addition to the risk of disease, the side-effects associated with HRT include mood changes, nausea, breakthrough vaginal bleeding and bloating, breast tenderness, concerns about breast,[3,4] and ovarian cancer,[5] gall bladder disease, and thromboembolic events.

Strong Medline warnings for estrogen now state: "Estrogen increases the risk that you will develop endometrial cancer (cancer of the lining of the uterus [womb]). The longer you take estrogen, the greater the risk that you will develop endometrial cancer. If you have not had a hysterectomy (surgery to remove the uterus), you might have been given another medication called a progestin to take with estrogen. This may decrease your risk of developing endometrial cancer, but may increase your risk of developing certain other health problems, including breast cancer."[6]

Concerns about safety and effectiveness are causing a retreat from the blanket use of HRT. An estimated 30%-45% of women who receive prescriptions for HRT will not have their prescriptions filled or will discontinue therapy within 12 months of initiation.[7]

Crucial link between Cholesterol, Magnesium and Hormones

It is impossible to consider estrogen and progesterone in isolation from other hormones and from precursors like cholesterol and magnesium. *All steroid hormones are created from cholesterol in a hormonal cascade.* Cholesterol is crucial for health and is the mother of hormones from the adrenal cortex, including cortisone, hydrocortisone, aldosterone, and DHEA. One of the most basic hormones and the first in the cascade is pregnenolone, which is converted into other hormones, including dehydroepiandrosterone (DHEA), progesterone, testosterone, and the various forms of estrogen. These hormones are interrelated, each performing a unique biological function.

DHEA is a steroid hormone produced by the adrenal gland and ovaries and converted to testosterone and estrogen. After being secreted by the adrenal glands, it circulates in the bloodstream as DHEA-Sulfate (DHEA-S) and is converted as needed into other hormones. Estrogens are also steroids.

Cholesterol cannot be synthesized without magnesium and cholesterol is a vital component of hormones. Aldosterone is one such hormone, and helps to control the balance of magnesium and other minerals in the body. Interestingly aldosterone needs magnesium to be produced and it also regulates magnesium's balance.[8] Women who suffer from premature menopause, or an early failure of the ovaries report that magnesium often helps fight the crashing fatigue that often comes at the beginning of premature or early menopause by boosting energy levels.[9]

The role that magnesium plays in the transmission of hormones (such as insulin, thyroid, estrogen, testosterone, DHEA, etc.), neurotransmitters (such as dopamine, catecholamines, serotonin, GABA, etc.), and mineral electrolytes is a strong one. Research concludes that it is magnesium status that controls cell membrane potential and through this means controls uptake and release of many hormones, nutrients and neurotransmitters.

"Magnesium," says Dr. Lewis B. Barnett, "is needed by the pituitary gland. The pituitary, sometimes called the miracle gland, takes instructions from the hypothalamus in the brain to which it is connected by a thin stalk, then transmits them through the body in the form of chemical messengers known as hormones. These hormones not only exert a direct influence of their own, but also trigger the production of other vital hormones elsewhere in the body. When the pituitary is not getting the magnesium it needs, it fails in its function of exercising a sort of thermostatic control over the adrenals which are thus allowed to overproduce adrenaline."

During perimenopause,[10] there are wide fluctuations in the hormones estrogen, testosterone, FSH, LH, and progesterone and it is these widely fluctuating hormone levels that can cause many problems, the least of which is hot flashes.

Estrogens are primarily responsible for the conversion of girls into sexually-mature women in the development of breasts, the further development of the uterus and vagina, broadening of the pelvis, growth of pubic and axillary hair and play a role in the increase in adipose (fat) tissue. They also participate in the monthly preparation of the body for a possible pregnancy and participate in pregnancy if it occurs. Estrogen is not one hormone, but many, and our bodies continue to make estrogens all of our lives. The adrenals, the fat tissues, and perhaps the uterus make estrogens.

Menopause is brought on by the natural decrease
in the body's production of estrogen and progesterone.

Estrogen causes a higher absorption and use of magnesium and zinc. Estrogen is normally associated with pregnancy. During pregnancy the body needs more minerals and estrogen takes care of the higher absorption. The estrogen enables a female to get just enough magnesium out of a low-magnesium diet. When the estrogen levels drop, the magnesium absorption drops and hypomagnesemia (magnesium deficiency) is the result. This can then cause a severe depression or diabetes or hypoglycemia or many other problems as both estrogen and magnesium levels drop through the years.[11]

The use of contraceptives, and estrogen replacement
therapies cause magnesium deficiencies.[12]

When you realize that more than three hundred types of tissues throughout the body have receptors for estrogen — which is to say that they're affected in some way by the hormone — it's not surprising that its decrease would cause physical changes. Estrogen affects the genital organs (vagina, vulva, and uterus), the urinary organs (bladder and urethra), breasts, skin, hair, mucous membranes, bones, heart and blood vessels, pelvic muscles, and the brain. It's the loss of estrogen to these organs that causes the ultimate changes of menopause, including dry skin and hair, incontinence and susceptibility to urinary tract infections, vaginal dryness, and, most important, the diseases osteoporosis and heart disease. These diseases are at the center of the controversy concerning menopause: Because estrogen plays a role in preventing these diseases, should you replace the estrogen lost at the time of menopause with a synthetic version?

Estrogens also have non-reproductive effects. They antagonize the effects of the parathyroid hormone, minimizing the loss of calcium from bones, and they promote blood clotting.[13] There are several forms of estrogen but the one most important for reproduction is estradiol, a substance secreted by the ovary. In addition to being responsible for the development of sexual characteristics in women, estrogen governs the monthly thickening of the endometrium and the quantity and quality of cervical and vaginal mucus so important to the successful passage of the sperm.

Magnesium is super critical to endrocrine function. Gonadotropin Releasing Hormone (GnRH) is a master hormone from the hypothalamus in the brain. It sparks the release of follicle stimulating hormone and luteinizing hormone from the pituitary gland, which in turn prompt production of estrogen and progesterone in the ovaries. Magnesium is involved in melatonin production and the circadian clocks in the human body. In particular, a deficiency of magnesium can impair the suprachiasmatic nucleus of the hypothalamus.[14] And balanced magnesium status is required to obtain efficiency of suprachiasmatic nuclei and the pineal gland.[15] Examinations of the sleep electroencephalogram (EEG) and of the endocrine system points to the involvement of the limbic–hypothalamus–pituitary–adrenocortical axis because magnesium affects all elements of this system. Magnesium has the property to reduce the release of adrenocorticotrophic hormone (ACTH) and to affect adrenocortical sensitivity to ACTH.

Hormone replacement therapy (HRT) is based on the incorrect assumption that your body becomes incapable of producing appropriate amounts of hormones simply because we reach a certain age. Your body does alter its hormone production as you pass through the stages of our life, but hormone problems are a function of how healthy you are, not how old you are.

Theresa Dale, ND

In today's age, with a staggering 68% of Americans not consuming the recommended daily intake of magnesium and more than 19% of Americans not consuming even half of the government's recommended daily intake of magnesium, we can easily see that magnesium impacts these life changes, the accompanying discomforts and can often reduce the problems and long term risks that occur.

Prior to menopause, estrogen plays a protective role in relation to heart disease, but as estrogen production diminishes, the risk of heart disease increases. Ten years after menopause, a woman has nearly the same risk as a man of dying of heart disease.[16]

Renowned magnesium researcher, Dr. Mildred Seelig points out that although there is no uniform agreement that estrogens lower serum magnesium levels, most of the evidence points in that direction. It is also possible that the paradoxical effects of estrogen on diseases of the cardiovascular system relate partially to its effects on magnesium distribution. It has been shown that serum magnesium falls with the cyclic increase in estrogen secretion. Since rats given estrogen showed decreased serum magnesium levels, without increased urinary magnesium output, and since the bone-magnesium increased, Goldsmith and Baumberger (1967) proposed that a shift of magnesium to the tissues was responsible for the estrogen-induced fall in serum magnesium.

The symptoms occurring during perimenopause can be severe and may correlate with naturally decreasing levels of DHEA which hit peak levels around the age of twenty and then decrease as we age.

Since DHEA is one of the primary bio-markers for aging, the long range effect of large doses of magnesium in a usable form is to significantly raise DHEA levels and thus produce true age reversal results. Dr. Norman Shealy, who is an expert on anti-aging, has done studies regarding magnesium and aging, refers to DHEA as the Master Hormone. He states that when produced at sufficient levels, DHEA will induce the production of all of the other hormones whose depletion can be associated with many symptoms of aging. He found that through the transdermal use of magnesium oil, women have reported complete abatement of menopausal symptoms and some have even returned to their menstrual cycle. This effect was found only when magnesium is applied through the skin, and not with oral products.

One of the major sexual impacts of decreased estrogen is a shrinking of the vagina and thinning of the vaginal walls, along with a loss of elasticity and decreased vaginal lubrication during sexual arousal. Some women experience only slight changes in sexual functioning, while others have dryness and pain with intercourse, or genital soreness for a few days after sexual activity, if they don't use a vaginal lubricant or take some form of hormone replacement. We have reports from some women using magnesium oil, that when sprayed in the vaginal area, lubrication is increased, vaginal dryness decreases, and sexual arousal is increased. Dr. Shealy confirms these findings from his clinical experience.

*It would seem from experimental studies
on animals that when one is low
on magnesium, small problems loom large,
even overpowering. Thus animals
deprived of magnesium suffer from super
excitability to such an extent that
they become hysterical at the sound of
small noises or the sight of shadows.*
J. I. Rodale

Premenstrual syndrome (PMS) is characterized by physical and emotional symptoms that develop following ovulation and decrease with the beginning of menstruation. These recurrent symptoms typically include anxiety, depression, irritability, fatigue, abdominal bloating, fluid retention in fingers and ankles, breast tenderness, altered sex drive, headache, and food cravings. The combination and severity of symptoms vary among women. The Office of Women's Health within the Department of Health and Human Services reports that as many as 75% of women experience some symptoms of premenstrual syndrome. This correlates quite closely with MIT's estimate that 68% of the population is deficient in magnesium.

*Natural supplementation with magnesium is highly
preferred over use of DHEA creams with their
many precautions and can relieve many of
these troublesome problem.*

The importance of balancing calcium with magnesium is noted by Dr. Christianne Northrup, who recommends a ratio of 1:1 between calcium and magnesium for PMS symptoms.[17] Magnesium supplementation has been shown, in double-blind trials, to be effective in relieving premenstrual symptoms. Dr. Melvyn R Werbach believes that even though many nutrients are implicated in the development of PMS, the borderline magnesium levels seen in PMS patients can explain most of the symptoms.[18] He notes that marginal deficiency of magnesium can deplete brain dopamine, impair estrogen metabolism, increase insulin secretion, and cause enlargement of the adrenal cortex (responsible for producing many hormones including sex hormones, stress hormones, and blood-sugar hormones).

"I think magnesium is the underrated all-star in terms of menopausal women," says Ann Louise Gittleman, PhD, pointing out it is not only good for bones, but it helps prevent heart disease and can keep you calm and help you sleep throughout the night. She recommends all women going through menopause take magnesium supplements along with Flax Seed.[19] Up to 80% of American women experience hot flashes during menopause while only 10% of Japanese women experience that symptom. Some researchers speculate that these differences may be due to differences in diet, lifestyle, and/or cultural attitudes toward aging.[20] But these suggested differences are vague and global in scope. In all likelihood the big difference is magnesium. Japanese women consume a large amount of sea vegetables of one kind or another all of which are extraordinarily high in magnesium.

Magnesium plays a critical role in a wide range of essential activities throughout the body, including many functions relevant to premenstrual changes experienced by some women. Magnesium is classed as "nature's tranquillizer" and so is vital in those aspects of the pre-menstrual symptoms which relate to anxiety, tension, etc. Women with PMS have been found to have lower levels of red blood cell magnesium than women who don't have symptoms and the supplementation of magnesium has been found to be extremely useful in alleviating many of the PMS symptoms and even more effective when taken with vitamin B_6 at the same time. A magnesium deficiency can cause blood vessels to go into spasms so if you suffer from menstrual migraines magnesium can be useful in preventing these spasms.

Magnesium is necessary for serotonin synthesis, which in turn is critical in mood regulation. Magnesium also appears to promote proper fluid balance, helping to ease the uncomfortable build up of excess fluid experienced by some women prior to menstruation. Inadequate magnesium levels have been found in women who experience premenstrual cravings and appetite changes.

A woman's menopause should not be seen as a pathologic endocrine deficiency disease because female hormones normally abate with advancing age as reproductive function comes to a halt. How and why this happens is a relative mystery to mainstream medicine but we can easily see how certain conditions will hasten and deepen the decline of the key hormones involved.

It is clear though that living without the protective effects of estrogen increases a woman's risk for developing serious medical conditions, including osteoporosis and cardiovascular disease. Women have every reason in the world to start supplementing their diets with large amounts of magnesium early in life, especially with magnesium chloride when applied transdermally. Though no one knows exactly why that form alone seems to provoke increases in DHEA levels, it probably has something to do with the penetration of the magnesium through the fat tissues.

Women should pay particular attention to adequate intakes of magnesium starting early on and supplement as necessary to assure adequate DHEA levels and better balanced hormone levels. Because women's issues are centered on hormonal balances it is vital to understand that the only way discovered so far to raise DHEA levels naturally is through transdermal application of magnesium chloride. Though magnesium chloride can be purchased in many pharmacies I highly recommend people experience a naturally made magnesium chloride that is a by-product of salt production. Below are some briefs on specific conditions related to menopause or menstruation where magnesium is shown to be of significant help.

Menstrual Migraine

Low magnesium levels may be a trigger for menstrual migraine. Mauskop et al reported a deficiency in ionized magnesium in 45% of attacks of menstrual migraine, while only 15% of non-menstrually related attacks had a deficiency. They also demonstrated that attacks associated with low ionized magnesium could be aborted by intravenous magnesium infusions. Facchinetti et al demonstrated that menstrual migraine could be prevented by administration of oral magnesium during the last 15 days of the menstrual cycle.

Menopause, Mood Disorders and Magnesium

Perimenopause and menopause related mood disorders cause significant distress to a large number of women. In the United States, one half of perimenopausal women will report feeling irritated or depressed.[21] Different studies have shown that a woman's risk for a first bout with depression rises sharply as she approaches menopause. "There is a subgroup of women who, for multiple reasons, may be more vulnerable," said Dr. Lee Cohen of Harvard Medical School, which followed 460 Boston-area women for six years.[22] Several studies,[23][24] show without doubt that there is a definite relation between magnesium deficiency and depression and that increasing our intake of magnesium can bring relief. Please see chapter on magnesium, violence and depression.

Osteoporosis

Each year over 300,000 women suffer a hip
fracture brought on by osteoporosis.
Within a year, one in five will die.

Magnesium plays a significant role in preventing Osteoporosis in the post menopausal period. Studies have shown that magnesium improves bone mineral density.[25] Without adequate magnesium, calcium cannot enter the bones.[26] Heavy metal exposure affects bone density. Although women with menopause may suffer from osteoporosis due to estrogen deficiency, bone fragility increases with increasing magnesium deficiency. High calcium intake is recommended for women with menopause, but adequate magnesium intake is necessary to lower dietary calcium/magnesium ratio, because the high ratio prompts blood coagulation. A group of menopausal women were given magnesium hydroxide to assess the effects of magnesium on bone density. At the end of the 2-year study, magnesium therapy appears to have prevented fractures and resulted in a significant increase in bone density.[27] The relationship between calcium and magnesium is dealt with extensively in the chapter on Calcium and Magnesium.

Magnesium and Hot Flashes

Many menopausal women suffer from heart palpitations associated with hot flashes. This can be helped by increasing your intake of magnesium. Magnesium plays a significant role in body temperature regulation.[28]

Studies in the use of therapeutic hypothermia have shown the efficacy of magnesium in lowering body temperatures. This supports the use of transdermal magnesium therapy for surface cooling by non-invasive methods.[29] Body temperature may be regulated by magnesium in two ways. One is through its central sedative effect on the hypothalamus and the second through its peripheral effect achieved by reducing the neuromuscular excitability.

Magnesium is lowered during hyperthermia due to its loss via sweat and magnesium diuresis.[30] Since we see that magnesium plays a significant role in regulation of blood sugars and regulation of body temperature, it makes good sense to use it for the treatment of vasomotor symptoms during menopause. Women can expect to find great improvement, more comfort, less mood disturbance and a smoother transition to post menopause. In addition Magnesium serves as a natural muscle relaxant, making it useful for relieving such symptoms as muscle cramping and anxiety.

References

1   See: www.healthfinder.gov/newsletters/heart042406.asp

2   In Greek, hormone means "to set in motion." Hormones are made by endocrine glands to control another part of the body. They require protein and fatty acids, cholesterol and magnesium to manufacture them. Many different hormones must be balanced one with another. This is done in at least two ways: (1) by the brain's information center, which monitors the state of the body, and (2) self-regulation as each gland detects chemical levels in the blood, giving "feedback" on the needs of the body. Glands may react by secreting one hormone to shut down the production or effects of another. Glands have the power to produce several different kinds of hormones at any time. The liver also has the power to control an overabundance of some hormones in the blood. Endocrine glands include the gonads, pineal, pituitary, thyroid, parathyroid, thymus and adrenals.

3   Colditz GA, Hankinson SE, Hunter DJ, Willett WC, Manson JE, Stampfer MJ, et al. The use of estrogens and progestins and the risk of breast cancer in postmenopausal women. *N Engl J Med* 1995;332:1589-93.

4   Collaborative Group on Hormonal Factors in Breast Cancer. Breast cancer and hormone replacement therapy. *Lancet* 1997;350:1047-59.

5   Garg PP, Kerlikowske K, Subak L, Grady D. Hormone replacement therapy and the risk of epithelial ovarian carcinoma: a meta-analysis. *Obstet Gynecol* 1998;92:472-9.

6   See: www.nlm.nih.gov/medlineplus/druginfo/medmaster/a682922.html

7   Hill DA, Weiss NS, LaCroix AZ. Adherence to postmenopausal hormone therapy during the year after the initial prescription. *Am J Obstet Gynecol* 2000;182:270-6

8   A deficiency in magnesium causes hyperplasia of the adrenal cortex, elevated aldosterone levels, and increased extracellular fluid volume. Aldosterone increases the urinary excretion of magnesium; hence, a positive feedback mechanism results, which is aggravated since there is no renal mechanism for conserving magnesium.

9   See: www.earlymenopause.com/9909.htm

10   Perimenopause is the naturally occurring transition period that takes place in women before the onset of menopause. It may begin as early as 35, even earlier for women who smoke. It is a temporary phase, typically lasting two to three years for most women, though for some it can last as long as 10 or 12 years. Women in perimenopause rank insomnia, irritability, and depressed mood among the most common complaints. Mental health is the most prevalent difficulty, not hot flashes. This stage of a women's life has not been talked about much, and a woman can find herself experiencing puzzling changes, and not know why. Studies have shown that in the perimenopause the incidence of negative changes was somewhat higher than in the postmenopause, the latter bringing relief of discomfort and a more positive mental outlook. Perimenopause terminates with the cessation of menstruation.

11   See: www.newtreatments.org/depression

12   Dahl, 1950; Nida and Broja, 1957; Goldsmith, 1963; Goldsmith et al., 1970; Goldsmith, 1971). The use of estrogen-containing oral contraceptives has been shown to reduce the serum levels of magnesium (in users versus nonusers) by 16% (Goldsmith et al., 1966), 28% (DeJorge et al., 1967), and by 27% and 33% (Goldsmith, 1971). Evaluation of different contraceptives suggests that it is the estrogen moiety that is responsible for the decrease in serum magnesium (Goldsmith and Goldsmith, 1966; Goldsmith et al., 1970, Goldsmith and Johnston, 1976/1980) although there are conflicting findings. So all the contraceptive pills, and hormone replacement estrogen preparations are probably decreasing women's magnesium levels too. Seelig, Mildred; For more information see: www.mgwater.com/Seelig/Magnesium-Deficiency-in-the-Pathogenesis-of-Disease/chapter5.shtml#toc5-1-4-3

13   See: users.rcn.com/jkimball.ma.ultranet/BiologyPages/S/SexHormones.html

14   See: www.ncbi.nlm.nih.gov/entrez/query.fcgi?cmd=Retrieve&db=pubmed&dopt=Abstract&list_uids=12635882

15   See: www.ncbi.nlm.nih.gov/entrez/query.fcgi?cmd=Retrieve&db=pubmed&dopt=Abstract&list_uids=12030424

16   Richard N. Ash MD; *Alternative Medicine and Health*; See: http://alternative-medicine-and-health.com/conditions/menopause.htm

17   Northrup, C. MD. *Women's Bodies, Women's Wisdom*. Judy Piatkus Publ. London, England, 1995.

18   Werbach, M. MD, *J Alt & Comp Med*. Feb. 1994;12(2).

19   *Take Control of Your Hot Flashes*: For more information, see: www.cbsnews.com/stories/2005/09/15/earlyshow/series/main848036.shtml

20   Reports differ but there has been some consensus that up to 80% of women in western societies such as Australia suffer from a myriad of physical and psychological difficulties at menopause (MacLennan, 1988). These include hot flushes, night sweats, vaginal dryness, loss of libido, palpitations, headaches, osteoporosis, depression and irritability (Walsh & Schiff, 1990). Interestingly, women in some non-western cultures appear to be significantly less affected by menopausal ills. For instance, Mayan women from South America (Beyene, 1986) and Rajput women in India (Kaufert, 1982) report no 'symptoms'. According to Lock et al (1988) Japanese women rarely mention hot flushes and the incidence of other problems such as backache and headache is low. It is therefore expected that due to the cross-cultural nature of the sample certain differences are likely to emerge with regard to physical, psychological and socio-cultural menopause experiences. *Women, body and society. Cross-cultural differences in menopause experiences*; Gabriella Berger & Eberhard Wenzel ; See: www.ldb.org/menopaus.htm

21   Obermeyer CM. Menopause across cultures: a review of the evidence. *Menopause* 2000;7:184-92.

22   Risk for new onset of depression during the menopausal transition: the Harvard study of moods and cycles. Cohen: *Arch Gen Psychiatry*. 2006 Apr;63(4):385-90.

23   Cerebrospinal fluid magnesium and calcium related to amine metabolites, diagnosis, and suicide attempts; Banki et al; *Biol Psychiatry*. 1985 Feb;20(2):163-71.

24   Treatment of severe mania with intravenous magnesium sulphate as a supplementary therapy. Heiden A et al; *Psychiatry Res.* 1999 Dec. 27; 89(3): 239-46

25   Institute of Medicine. Food and Nutrition Board. Dietary Reference Intakes: Calcium, Phosphorus, Magnesium, vitamin D and Fluoride. National Academy Press. Washington, DC, 1999

26   Aging and magnesium; Saito N, Nishivama S; *Clin Calcium.* 2005 Nov;15(11):29-36.

27   Magnesium supplementation and osteoporosis. Seijka Je, Weaver; *Nutr Rev.* 1995 Mar;53(3):71-4

28   How significant is magnesium in thermoregulation? *J Basic Clin Physiol Pharmacol.* 1998;9(1):73-85. PMID: 9793804 [PubMed - indexed for MEDLINE]

29   Therapeutic hypothermia shows promise as a treatment for acute stroke. Surface cooling techniques are being developed but, although noninvasive, they typically achieve slower cooling rates than endovascular methods. We assessed the hypothesis that the addition of intravenous $MgSO_4$ to an antishivering pharmacological regimen increases the cooling rate when using a surface cooling technique. Subjects who received $MgSO_4$ had significantly higher mean comfort scores than those who did not (48+/-15 versus 38+/-12; P<0.001). CONCLUSIONS: Administration of intravenous $MgSO_4$ increases the cooling rate and comfort when using a surface cooling technique. Magnesium sulfate increases the rate of hypothermia via surface cooling and improves comfort. *Stroke.* 2004 Oct;35(10):2331-4. Epub 2004 Aug 19. PMID: 15322301 [PubMed - indexed for MEDLINE] http://ncbi.nlm.nih.gov/entrez/query.fcgi?cmd= Retrieve&db=pubmed&dopt=Abstract&list_uids=15322301&itool=iconfft&query_ hl=4&itool=pubmed_docsum

30   See: http://72.14.203.104/search?q=cache:2Lsm-2ZiMbQJ:www.profmagnesium. com/PDF%27S/Mg%2520in%2520Thermoregulation%25201997.pdf+magnesium+an d+thermoregulation&hl=en&gl=us&ct=clnk&cd=5

21

||||||||||||||||||||||||||||||||||||||||||||||||||||||||||||||||||||||

Natural Relief for Chronic Pain Sufferers

Americans' intake of magnesium dropped 50% in the last century, and the consequences are quite painful.

When I received the following account from my research assistant Claudia French, who is an RN in an acute care psychiatric hospital, I realized that we should address the issue of magnesium and pain more directly.

Yesterday I witnessed one of the most amazing benefits of transdermal magnesium I have seen. I work with another RN who is afflicted with arthritis, especially in her hands, and frequent muscle cramping/spasms in her legs. She has been using magnesium but became lax. Before leaving for work yesterday I received a phone call from her begging me to please bring with me some magnesium oil, as her hands were so cramped up and painful that she could barely stand to continue working.

When I got there, her hands and fingers were very contorted in spasm. Her fingers were curled up and stiff and her legs were cramping badly. She reported they had been this way all day, and the pain was driving her to tears. She immediately slathered the magnesium oil all over her hands. We were in report and she wanted it on her hands right away so the entire nursing staff watched and within 5 minutes you could visibly see her fingers extend back to normal and the finger movement return. We could literally see the relaxation taking place. It was simply amazing. Within minutes her hands were completely relaxed and functional again and stayed that way the remainder of the evening. She also applied the magnesium to her legs and found relief.

About 30 minutes after applying the oil, she held up her hands for everyone to see, and showed us the arthritic nodules on some fingers. She described how painful these always are to touch. But she poked and prodded them telling us how there was no pain now. She was able to continue working and doing the extensive writing that is a large part of our work without any further discomfort.

Pain relief and muscle relaxation for people with arthritis and muscle cramping is an important and significant benefit of magnesium oil. The rapid relief, visible to us all was really amazing! The following day she reported that she'd gotten the first restful night of sleep in many days. The pain was not waking her up.

Dr. Linda Rapson, who specializes in treating chronic pain, believes that about 70% of her patients who complain of muscle pain, cramps and fatigue are showing signs of magnesium deficiency. "Virtually all of them improve when I put them on magnesium," says Rapson, who runs a busy Toronto pain clinic. "It may sound too good to be true, but it's a fact." She's seen the mineral work in those with fibromyalgia, migraines and constipation. "The scientific community should take a good hard look at this."[1]

Lynne Suo is one of Dr. Rapson's patients. She had been using painkillers and steroids for years to try to ease the pain of her arthritis and fibromyalgia. Dr. Rapson started her on 675 units of magnesium a day. Within days, Suo called Dr. Rapson to report a surprising change. "I went from being in constant pain almost throughout the day and night to having moments of pain. And for me that was a huge improvement," says Suo, a former college English teacher. She dismisses suggestions that the change is a placebo effect. "I was not one day without pain and now I don't have to take heavy pain medication," she reports.

We know that a lack of magnesium underlies our epidemic of heart disease, high blood pressure, diabetes and osteoporosis. Minus magnesium, hearts beat irregularly; arteries stiffen, constrict and clog; blood pressure rises; blood tends to clot; muscles spasm; insulin grows weaker and blood sugar jumps; bones lose strength; and pain signals *intensify*.

"Many people needlessly suffer pain - including fibromyalgia, migraines and muscle cramps — because they do not get enough magnesium," says Mildred Seelig, M.D., a leading magnesium researcher at the University of North Carolina. The problem is exacerbated when they load up on calcium, thinking it will help, when in fact, an overabundance of calcium flushes magnesium out of cells, compromising the effectiveness of *both* minerals. Prescription medications, such as antidepressants, tranquilizers and pain medications, only treat the symptoms. Magnesium treats the symptoms while it simultaneously addresses the cause of much of the pain and disease we experience. In fact, it could be surmised that pain and disease is one residual effect of magnesium deficiency, or a mineral imbalance.

Intravenous infusions of magnesium sulphate
reduced intra and postoperative analgesic
consumption in a recent clinical study.[2]

Why is magnesium not promoted by doctors as the pain relief medicine it truly is? "It's not taught in medical schools," said the late Dr. Mildred Seelig, a former professor of nutrition at the University of North Carolina. She studied magnesium for years and urged doctors and patients not to overlook this essential mineral. "It's not being promoted because pharmaceutical companies don't make money selling magnesium," said Dr. Seelig. "So there is no big push to get magnesium understood and taken by the average (North) American."

Experimental systemic and intrathecal
injection of magnesium suppressed neuropathic
pain responses via a spinal site of action in rats
with chronic constriction injury of the
sciatic and saphenous nerves.[3]

Mention magnesium, and many people conjure up images of a hard, silvery alloy used to fashion parts for aircraft and automobiles, or machinery that needs to resist corrosion. Mention it as a pain reliever second to none and people will scratch their head and wonder what's wrong with you. But medical scientists from the Department of Anaesthetics and Pain Management, University Hospital Lewisham, London, think the beneficial effect of magnesium in terms of pain management may result from the physiological action of magnesium as a *non-competitive antagonist* of the NMDA-receptor. Thus intravenous or transdermal magnesium treatments will increase the Mg++ concentration gradient between the extracellular fluid and the cell membranes, causing a block of the NMDA-receptor and subsequent pain relief.

The N-methyl-D-aspartate (NMDA) receptor plays an important role in the mechanisms underlying central sensitization in the spinal cord, which is critically important for the establishment of several chronic neuropathic pain states.[4] *In its inactive state the NMDA receptor is blocked by the presence of a centrally positioned magnesium ion.* Given that simple piece of information alone, it becomes easy to discern how having sub-optimum levels of magnesium in the system would result in one experiencing higher levels of pain. Furthermore, pain medications are not substitutes for magnesium. Pain signals may be temporarily suppressed, but the condition that caused it goes unaddressed.

NMDA receptor activation and the concomitant release of pro-pain substances such as substance P (SP), nerve growth factor (NGF), brain derived nerve factor (BDNF), and nitric oxide is believed to drive the process of central sensitization. A recent text stated that the 'recruitment of the NMDA receptor appears to be the pivotal event in increasing the sensitivity of the nociceptive (pain) spinal circuits to (painful stimuli)". This is believed to occur when repetitive pain stimuli knock the magnesium block off the NMDA receptors (Staud 2004).

NMDA receptors form ion channels at the excitatory synapses of nerves. Ordinarily these ion channels are a) closed and b) blocked by extracellular magnesium ions. Two things need to happen for these ion channels to open, first adequate glutamate needs to be present to open the channel and the membrane of the cell needs to become depolarized in order to flush the magnesium out. The negatively charged cell attracts magnesium ions which then close the NMDA ion channels. If the cell loses its negative charge, the magnesium ions will drift away.[5]

The loss of the magnesium ions allows sodium, potassium and most importantly calcium ions into the cell. Once inside the nerve cell the calcium ions activate signaling pathways that call for the production of glutamate, substance P and other pro-pain factors. The neurons in the spinal cord then send a message to the part of the brain called the thalamus which processes it and amplifies the pain response. Prolonged periods of NMDA ion channel activation and the high intracellular Ca^{2+} levels that accompany it can, by triggering the apoptotic process (cell suicide), result in nerve cell death.

Interestingly researchers have investigated the effect of the combination of magnesium and morphine in experimental models of chronic and tonic pain. They found that magnesium alone induced a significant antihyperalgesic effect in mononeuropathic and diabetic rats after a cumulative dose of 90 mg/kg. Magnesium was found to amplify the analgesic effect of low-dose morphine in conditions of sustained pain. Considering the good tolerability of magnesium, these findings have clinical applications in neuropathic and persistent pain.[6]

The key receptor in the central nervous system
for pain modulation is the NMDA receptor.

When these receptors are made more excitable, which they are in conditions of magnesium deficiency, the pain experience is upgraded. In fact all receptors are functionally dependant on the status of their voltage dependant channels, or gates. These gates are typically specific for minerals such as calcium, magnesium, potassium or sodium. Here is the exact point that reinforces the idea that mineral balance and sufficiency, especially in magnesium, are keys to optimum health, and how their absence or imbalance causes us to be in pain, and therefore, seek ineffective pain management regimens.

Several neuropathic disorders including diabetic peripheral neuropathy, postherpetic neuralgia, Reflex Sympathetic Dstrophy (Complex Regional Pain Syndrome), post surgical/traumatic neuropathy, toxic and idiopathic neuropathies share similar pain perception pathways that will be helped by magnesium supplementation delivered locally and systemically transdermally.

The transdermal approach, via baths, footbaths, or topical application, has the unique added advantage of allowing the body to regulate and administer the magnesium where it is needed, and in what amounts. The body will not "overdose" itself, or work against its best interests. When it has enough — for example, from taking a bath where an amount of magnesium chloride has been added — it would simply stop absorbing more. All other methods of supplementation involve educated guesses.

In the 1990's cardiovascular biologist Dr. Burton M. Altura of the State University of New York Health Science Center at Brooklyn witnessed a therapeutic benefit of magnesium in acute symptoms, such as headache pain. Altura administered a solution containing 1g of magnesium sulfate intravenously to 40 patients who visited a headache clinic in the throes of moderate to severe pain. They treated not only migraine sufferers but also persons with cluster headaches and chronic daily headaches.

Within 15 minutes, 32 of the men and women — 80% — experienced relief. Though the headache may not have vanished, the pain lessened by at least 50%. In 18 of these individuals, the pain relief lasted at least 24 hours. Blood tests before treatment confirmed that all but four in this latter group had ionized magnesium concentrations that were lower than the average in a related group of pain free individuals. "All nine patients with cluster headaches had their acute headache aborted by magnesium therapy." Migraine sufferers who responded to the treatment experienced a complete alleviation of their current symptoms, including sensitivity to lights and sound. Subsequent studies of additional migraine patients have confirmed a common pattern, Altura says. "Those patients where ionized magnesium in the brain or blood is low will respond to intravenous magnesium very quickly and dramatically."

The combination of heat and magnesium chloride increases circulation and waste removal. The therapeutic effect of magnesium baths is to draw inflammation out of the muscles and joints. A whole new world of pain management will be realized when doctors and patients find out that magnesium chloride from natural sources is available for topical use and that the potential for pain relief is enormous.

Magnesium chloride, when applied directly to the skin is transdermally absorbed and has an almost immediate effect on chronic pain. Transdermal magnesium therapy is also ideal for athletes who need high levels of magnesium. Oral magnesium is much less effective than transdermal magnesium in the treatment of injuries and tired worn out muscles. Perhaps the biggest difference between oral and transdermal supplementation of magnesium is seen in the area of pain management.

A friend of mine Dr. David I. Minkoff called me recently complaining about tired and sore aching muscles from the strenuous athletic training he, a 58 year old, was subjecting himself to. He was putting in 20 hours of training/wk for competition in his 32nd Ironman triathalon. He was using the magnesium oil but just a little bit of it. I told him to buy a gallon and dump whatever he had into his bath right away.

> "I did the magnesium soak two days ago as you said with 4 ozs. The next morning I was better. Yesterday I did a 101 mile bike ride up a 6200 foot mountain. Was out in the heat (90° F.) for 7½ hours and then ran two miles when I got home. I am usually cramped up when I get home after a day like this and feeling pretty done in. But no cramps. Did another soak after the ride and run (I used more as my gallon arrived) and today I am not sore at all. I should be limping around. I did an easy 2.5 mile ocean swim and did another 6 oz. in the tub, then got a massage and my body is feeling good."

Transdermal magnesium therapy is an
ideal pain management treatment system.

According to Dr. Cathy Wong, a German study found that mineral supplements increased intracellular magnesium levels by 11% and was associated with a reduction in pain symptoms in 76 out of 82 people with chronic low back pain.[7] London researchers provide strong evidence that magnesium sulphate produces pain relief in patients with PHN[8], a neuropathic pain condition.[9]

Magnesium deficiency is common in people with depression and chronic pain. Major depression is thought to be four times greater in people with chronic back pain than in the general population (Sullivan, Reesor, Mikail & Fisher, 1992). It has been found that the rate of major depression increased in a linear fashion with greater pain severity (Currie and Wang, 2004). Thus magnesium, when applied transdermally, since it is good for both chronic back pain and depression separately, is the nutrition/medicine of choice for treating these sufferers.

Natural pain relief with transdermal magnesium chloride therapy is safe and effective. Magnesium may act as good opiate alternative for many who become easily dependent on narcotic painkillers whose ill affects are permeating the population and even now are getting into the hands of our children. Chronic pain is often treated with opiates, and the effects are terrific as long as they last, but tolerance develops rapidly, and increasing doses are required to produce the same pain relieving effect, further adding to the difficulty in treating pain.

Most pain medications are not safe; even the over the counter pain medications hold unforeseen dangers. Despite more than a decade's worth of research showing that *taking too much acet-aminophen can ruin the liver*, the number of severe, unintentional poisonings from the drug is on the rise, a new study reports.[10] The drug, acetaminophen, which was approved by the FDA in 1951, is best known under the brand name Tylenol. Compounds containing acetaminophen include Excedrin, Midol Teen Formula, Theraflu, Alka-Seltzer Plus Cold Medicine, and NyQuil Cold and <u>Flu</u>, as well as other over-the-counter drugs and many prescription nar-cotics, like Vicodin and Percocet.

Dr. William Lee, a liver specialist at the University of Texas Southwestern Medical Center in Dallas said he was disturbed by a pattern: "that acetaminophen is always billed as the one to reach to for safety, probably even more so now, with other pain relievers pulled from the market." "It's extremely frustrating to see people come into the hospital who felt fine several days ago, but now need a new liver," said Dr. Tim Davern, a gastroenterologist with the liver transplant program of the University of California at San Francisco. "Most had no idea that what they were taking could have that sort of effect."

What better way to reduce or eliminate pain then by sim-ply taking a bath or rubbing magnesium chloride in liquid form directly onto the skin or affected area of the body. From the pain of sports injuries to low back pain and sciatica, headaches, relief from kidney stones, the pain of restless legs, arthritic pain, and just about every painful condition imaginable will in all likely hood benefit from magnesium chloride applied topically.

What is essential to remember about treating pain with mag-nesium is that it treats both the symptom and the cause of pain. Meaning the cause of the pain can often be traced back to a magnesium deficiency. Transdermally applied magnesium chloride easily belongs in the middle of every pain program and can be used in conjunction with all other pain medications.

References

1   See: www.ctv.ca/servlet/ArticleNews/story/CTVNews/20020923/favaro_magnesium020923/CTVNewsAt11/story/

2   Koinig H, Wallner T, Marhofer P, Andel H, Horauf K, Mayer N. Magnesium sulphate reduces intra-and postoperative analgesic requirements. *Anest Analg* 1998; 87(1):206-210.

3   Xiao WH, Bennett GJ. Magnesium suppresses neuropathic pain responses in rats via a spinal site of action. *Brain Res* 1994; 666(2):168-172.

4   Woolf CJ, Thompson WN. The induction and maintenance of central sensitization is dependent on N-methyl-D-aspatic acid receptor activation: implications for the treatment of post-injury pain hypersensitivity states. *Pain* 1991; 44:298-9.

5   Tanaka, M., Sadato, N., Mizuno, K., Sasabe, T., Tanabe, H., Saito, D., Onoe, H., Kuratsune, H. and Y. Watanbe, Y. 2006. Reduced responsiveness is an essential feature of chronic fatigue syndrome: an fMRI study. *BMC Neurology.* 6: 9 doi:1186/1471-2377-6-9

6   Laboratoire de Pharmacologie Medicale, Faculte de Medecine, Clermont-Ferrand, France. *Anesthesiology* 2002 Mar;96(3):627-32

7   For more information, go to: http://altmedicine.about.com/od/chronicpain/a/back_pain_2.htm

8   Post herpetic neuralgia (PHN) is a complication of acute herpes zoster infection (HZ), characterized by severe constant pain and disturbances of the sensory nervous system in the skin area initially affected by the infection.

9   Brill, S.; Sedgwick, P.; Hamann, W. Magnesium Relieves Pain in Posterpetic Neuralgia Department of Anaesthetics & Pain Management, University Hospital Lewisham, London, United Kingdom;

10   Source: The New York Times (11/29/2005).

22

|||||||||||||||||||||||||||||||||||||||||||||||||||||||||||||||||||

Testing and Estimating Magnesium Levels

I f you doubt anything about your health you should doubt your cellular magnesium levels. Though you will still see many statements indicating that magnesium deficiency is rare, don't believe it for a moment. Even some of the sites that offer magnesium testing say this but it is based on the fact that the most popular magnesium test is for blood serum levels. Since magnesium is mainly an intracellular ion, measurement of serum total magnesium is an inaccurate index of intracellular or total magnesium stores. This means that although your serum levels may be maintained within normal limits, there could be a deficiency in tissues that is not being detected. Blood serum tests will almost always show normal no matter what the magnesium levels are in the cells because the body tightly controls the level of magnesium in the blood because if that level falls a heart attack is never that far away.

*Serum levels of magnesium must be kept within
a tight range, or the heart stops. Therefore serum levels
are maintained at the expense of levels inside cells.*
                                                Dr. Sarah Mayhill

When it comes to magnesium it is foolish to place all our faith in tests, especially the blood serum test. Less than 1% of our body's total magnesium can be measured in our blood; the rest is found in the cells and tissues of the body where it is needed for crucial cell processes. It's impossible to make a diagnosis about magnesium levels this way, for magnesium in the blood does not correlate with magnesium levels in the rest of the body. In fact, when we are under stress, our body dumps magnesium into the blood giving the mistaken appearance of normal levels even when the rest of the body is terribly deficient. This is just one more reason to be wary of doctors, clinics and hospitals that continue to rely on a virtually irrelevant magnesium test.

Many allopathic doctors implore you to obtain an accurate diagnosis before trying to find a cure. They remind you that many diseases and conditions share common symptoms: if you treat yourself for the wrong illness or a specific symptom of a complex disease, you may delay legitimate treatment of a serious underlying problem. The implication is that the greatest danger in self-treatment may be self-diagnosis. If you do not know what you really have, you can not treat it!

This is a lot of nonsense and mirrors a medical system that has lost contact with reality. When it comes to magnesium and its deficiency, we can safely assume that we are deficient to one degree or another, even if we feel perfectly healthy. The chances today of running through life with a full tank of magnesium without any kind of supplementation are about zero. Perhaps the five percent of the population who eat organically and supplement with super foods like spirulina, wheat grass juice, and sea vegetables like the Japanese eat, will not be deficient. However, if you are presently ill, or chronically so, the chances are probably less than zero that you are not deficient.

In 1936, testimony was put before the American Congress attesting that the food we produce and eat was devoid of basic nutrients. Over 70 years later, the situation is far worse and the basic picture is frightening. This is being reflected in the explosive growth of chronic diseases in the old and young.

No man, woman or child today can eat enough fruits and vegetables to supply their bodies with enough magnesium for perfect health. There has been a gradual decline of dietary magnesium in the United States, from a high of 500 mg/day at the turn of the last century to barely 175-225 mg/day today.[1] Both MIT and the National Academy of Sciences have determined that there are vast deficiencies in the American population. When you consider that the Recommended Daily Allowance (RDA) is severely underestimated for magnesium, it becomes clear that almost 100% of the population would be magnesium deficient.

Still, the most common question people who become aware of magnesium want to know is, "How do I know if my body is magnesium deficient?" In addition to the above factors one should also consider the fact that if you are under mild to moderate stress caused by physical or psychological disease, physical injury, athletic exertion, or emotional upheaval, your requirements for magnesium escalate.[2,3] Today, who is not under increasing stress?

Actually, one of the best ways to determine if you are magnesium deficient is to begin applying magnesium chloride transdermally to the body in low doses and to see within days, if not hours, if you feel better. Often conditions that are marked by pain will show improvement almost immediately as the magnesium ions penetrate through the skin. Though as we see from the testimonial below if one is intensely deficient one might feel discomfort when applying the magnesium chloride at first.

> "When I first began applying magnesium directly onto my arm pit, it would burn intensely for a period of time, and then stop. However, I was also amazed to discover that I emitted no body odor after doing this, even when I'd sweat. So I decided to endure the pain for the benefits that I knew were happening. But the pain was indeed intense, and appeared to last for a longer duration, to the extent that I finally dialed back the concentration, mixing magnesium with energized water in a 50/50 solution. This lessened the pain, but didn't end it."

When one is underwater too long and comes up for air, that first deep breath is deeply appreciated by the total organism that we are. In reality it's not that much different with magnesium. After years if not decades of deficiency, the body responds quite powerfully to the systemic application of magnesium. When taking a bath in magnesium chloride or when applying it directly to the skin the body receives the magnesium in a way that is only approached by intravenous treatments.

One would not wait for a doctor's permission or a long and expensive diagnostic process before taking that first life saving breath. One should not wait for anyone to add more magnesium into ones body. In today's toxic world the most basic condition that leads to disease is the "one/two punch" of dietary deficiency faced with rising levels of toxicity in the body. Everyone is exposed to dangerous chemicals that penetrate and accumulate in the body. When our levels of magnesium, and other vital minerals and nutrients are low, the body just cannot handle the increasing toxicity. This is too simple for doctors addicted to the pharmaceutical companies that actually want you to increase the overall toxicity of your body by taking more of their toxic drugs.

If you still want to test for magnesium deficiency, there are other tests besides the blood serum test:

Total Red Cell Magnesium. The results of this test are less variable than serum measurements, but it may still not adequately reflect total body magnesium status in health and disease. This test measures the amount of the mineral magnesium inside the red blood cells. However, total red blood cell magnesium levels are not as accurate a measurement of tissue levels as the ionized magnesium test.

Serum Ionized Magnesium. The blood ionized magnesium test correlates well with intracellular free magnesium levels. Dr. Carolyn Dean favors the blood ionized magnesium test saying it is a very refined procedure backed by results on many thousands of patients and information about it has been published in prestigious journals.

Intracellular Free Magnesium. Nuclear magnetic resonance (NMR) spectroscopy is another way to test for magnesium, but it is impractical because of the cost and the lack of routine availability.

Sublingual Magnesium Assay. The Buccal cell smear test or 'Exatest' is a safe, non-invasive test that accurately measures the minerals inside cells. This is a test used, for example, during cardiac surgery to determine cellular magnesium levels. A doctor painlessly collects a sample from under your tongue and affixes it to a slide. The slide is then sent to IntraCellular Diagnostics, Inc. for analysis.

Magnesium Loading Test. This test measures urinary magnesium excretion in response to a loading dose of magnesium. Although inconvenient to perform, this test has successfully identified individuals with even mild degrees of magnesium deficiency. It has been considered an accurate test when renal function is normal.

The ultimate and most dependable indicator of magnesium deficiency is your health. If your body and mind are functioning perfectly; if you have the desire, will, and energy to pursue your life dreams; if you are not saddled by aches and pains of the head, heart, or mind, then you are most likely not magnesium deficient.

Yet, if the Standard American Diet is your primary source of nutrition you can be assured that the magnesium stores in your body will steadily decline. If you drink, rest assured that will further push you into magnesium deficiency and this will be reflected in your state of health.[4] So if any of these factors are true, then you need to replenish your magnesium stores, irrespective of what any tests say. Remember that the ultimate litmus test for optimum health, is optimum health, and optimum health cannot be sustained without sufficient magnesium stores.

References

1   Altura, BM, "Introduction:importance of Mg in physiology and medicine and the need for ion selective electrodes." *Scand J Clin Lab Invest Suppl*, vol. 217, pp 5-9, 1994

2   Seelig MS, "The requirement of magnesium by the normal adult." *Am J Clin Nutr*, vol 14, pp.342-390, 1964

3   Seelig MS, "Magnesium requirements in human nutrition." *Magnes Bull*, vol 3 (1A), pp. 26-47, 1981

4   Both acute intake of alcohol and chronic alcoholism can have detrimental effects on magnesium nutriture. This review by Dr Richard S Rivlin of New York Hospital-Cornell Medical Center summarizes the evidence linking alcohol with magnesium deficiency and describes how this effect may contribute to the complications of alcoholism, including delirium tremens and liver cancer. Many studies have shown that the state of chronic alcoholism is associated with magnesium deficiency. This deficiency has multiple causes, including the increase in magnesium excretion in response to and the anorexia, erratic eating habits, malabsorption (especially fat malabsorption), and diarrhea commonly found in alcoholics. Some procedures used in the treatment of severe alcoholics may exacerbate magnesium deficiency. For example, because zinc deficiency is well recognized as a complication of alcoholism, patients are often given high-dose zinc supplements. However, the administration of large amounts of zinc leads to an increase in the excretion of magnesium. The use of diuretic drugs to correct fluid and electrolyte problems and the prolonged administration of intravenous fluids that do not contain magnesium may also contribute to magnesium depletion in patients being treated for severe alcoholism. **Review**: alcohol, magnesium, and cancer - liver cancer - adapted from *J American College Nutrition*, October 1994 issue. *Nutrition Research Newsletter*, Oct. 1994 For more info, see www.findarticles.com/p/articles/mi_m0887/is_n10v13/ai_15882994.

23

||||||||||||||||||||||||||||||||||||||||||||||||||||||||||||||||||||||

Warnings and Contraindications

Toxic symptoms from increased magnesium intake are not common because the body eliminates excess amounts unless there are serious problems with kidney function. Magnesium excess sometimes occurs when magnesium is supplemented as a medication (intravenously) because adding magnesium in very large doses, in isolation from other nutrients, can cause harmful effects on the body. In reality, problems with magnesium supplementation usually occurs when the magnesium in the IV is given too rapidly and in too high of a dose or both.

There is the balance of calcium to magnesium to be kept in the range of 1:1 to 2:1. If more magnesium than calcium is taken then you are going to upset your calcium balance. This is not an issue though for people whose dairy intake is high. Most people today are getting too much calcium and not enough magnesium.

The ratio of minerals and vitamins to each other is important. Scientists from the University of Helsinki said, "The present average sodium intakes, approximately 3000-4500 mg/day in various industrialized populations, are very high, that is, 2-3-fold in comparison with the current Dietary Reference Intake (DRI) of 1500 mg.

The sodium intakes markedly exceed even the level of 2500 mg, which has been recently given as the maximum level of daily intake that is likely to pose no risk of adverse effects on blood pressure or otherwise.

By contrast, the present average potassium, calcium, and magnesium intakes are remarkably lower than the recommended intake levels (DRI). In the USA, for example, the average intake of these mineral nutrients is only 35-50% of the recommended intakes. There is convincing evidence, which indicates that this imbalance — i.e., the high intake of sodium on one hand and the low intakes of potassium, calcium, and magnesium on the other hand — produce and maintain elevated blood pressure in a significant portion of the population.

Decreased intakes of sodium alone, and increased intakes of potassium, calcium, and magnesium each alone decrease elevated blood pressure. A combination of all these factors, that is, decrease of sodium, and increase of potassium, calcium, and magnesium intakes, which are characteristic of the so-called Dietary Approaches to Stop Hypertension diets, has an excellent blood pressure lowering effect."[1]

In isolation and in too high a quantity, anything can become a problem. There is a balance needed between minerals, trace elements and large amounts of magnesium used to treat disorders. Spirulina is offered as the ideal complement to transdermal magnesium chloride therapy for it is a potent medicine in its own right and is another gift from the Waters of Life.

Spirulina, which is high in chlorophyll, is probably the most potent food on planet earth and provides a complete list of all the minerals and trace elements as well as amino acids and fatty acids we need to sustain life. Anything that has chlorophyll has magnesium, since it is the center of the chlorophyll molecule.

Some people and especially children might develop a rash from using the magnesium oil when applied directly to the skin. Many children, if the magnesium oil is used at full strength, will feel a burning or stinging and this can be painful and if this happens the oil should be washed off quickly. In such cases you need to dilute the magnesium oil 50/50 with distilled or mineral water and when the body acclimates to the magnesium one can then build up to the full concentration. A rule of thumb about dosage: It is always a good idea to start with low dose and work one's way gradually to higher doses. Whenever any kind of uncomfortable reaction occurs this is a sign to lower the dosage or concentration.

Magnesium is regulated and excreted primarily by the kidneys where various ATPase enzymes are responsible for maintaining homeostasis.[2] However hypermagnesemia can also occur in people with hypothyroidism, those using magnesium containing medications such as antacids, laxatives, cathartics, and in those with certain types of gastrointestinal disorders, such as colitis, gastroenteritis and gastric dilation, which may cause an increased absorption of magnesium.

Risk of magnesium toxicity is usually related to severe renal insufficiency, when the kidney loses the ability to remove excess magnesium. Individuals with impaired kidney function are at higher risk for adverse effects from magnesium supplementation and people with severe renal insufficiency should avoid magnesium supplementation or approach very carefully starting with very low dosages.

Everyone needs magnesium and its deficiency itself can lead to problems in any part of the body including the kidneys. Magnesium is essential for life, as is air, and in no situation can the body live without it. Magnesium supplementation in children with dehydration or renal failure is also contraindicated so before beginning any kind of magnesium treatment any dehydration needs to be addressed.

Signs of excess magnesium symptoms can be very subtle and can occur with long term use of magnesium supplements and laxatives. The symptoms can be similar to magnesium deficiency and include: changes in mental status, nausea, diarrhea, appetite loss, muscle weakness, difficulty breathing, extremely low blood pressure, and irregular heartbeat. Though extremely rare severe magnesium intoxication is manifested by a sharp drop in blood pressure and respiratory paralysis. Disappearance of the patellar reflex is a useful clinical sign to detect the onset of magnesium intoxication. In the event of over dosage, artificial ventilation must be provided until a calcium salt can be injected by IV to antagonize the effects of magnesium.

The most common cause of hypermagnesemia is renal failure. Other causes include the following:

- Excessive intake
- Lithium therapy
- Hypothyroidism
- Addison disease
- Familial hypocalciuric hypercalcemia
- Milk alkali syndrome
- Depression

Most adverse effects of parenterally administered magnesium (intravenous) are usually the result of magnesium intoxication. These include flushing, sweating, hypotension, depressed reflexes, flaccid paralysis, hypothermia, circulatory collapse, cardiac and CNS depression proceeding to respiratory paralysis. Hypocalcemia, with signs of tetany secondary to magnesium sulfate therapy for eclampsia, has been reported.

Intravenous administration of magnesium could accentuate muscle relaxation and collapse the respiratory muscles if given too rapidly or in too high a dosage. Patients with excessively slow heart rates should also be careful because slow hearts can be made even slower, as magnesium relaxes the heart. When there is an obstruction in the bowel additional caution is required because the main route of elimination of oral magnesium is through the bowel.

Magnesium supplementation is known to interact with many different pharmaceutical drugs and it is wise to know what these are when treating our patients. Certain drugs will increase the loss of magnesium in urine. Thus, taking these medications for long periods of time may contribute to magnesium depletion. On the other hand many antacids and laxatives contain magnesium. When frequently taken in large doses, these drugs can inadvertently lead to excessive magnesium consumption and hypermagnesemia, which refers to elevated levels of magnesium in blood.

Some recommendations on dosing related to medications when used with magnesium are:

Doxycycline

Magnesium may make doxycycline less effective. Take magnesium supplements 1 to 3 hours before or after ingesting doxycycline.

Minocycline

Magnesium may make minocycline less effective. Take magnesium supplements 1 to 3 hours before or after ingesting minocycline.

Tetracycline Hydrochloride

Magnesium may make tetracycline less effective. Take magnesium supplements 1 to 3 hours before or after ingesting tetracycline.

The following diabetes medicines:

Glipizide (Glucotrol®) and Glyburide (Micronase, Glynase, Diabeta). Taking magnesium and either Glipizide or Glyburide together may further lower blood sugar leading to blurred vision, tremor (shaking), hunger, sweating, headache, skipped heart beats, confusion, nervousness and extreme tiredness. Magnesium (also commonly found in antacids), may increase the absorption of glipizide and glyburide, medications used to control blood sugar levels. Ultimately, this may prove to allow for reduction in the dosage of those medications.[3]

The Magnesium Research Institute says that the drug Neurontin binds magnesium in the GI tract and results in a malabsorption of both oral magnesium and Neurontin (PDR says 24%). Interaction with Neurontin is important to note, because it is an anti-seizure medication, and also used off label, frequently, as a mood stabilizer and behavioral drug, in addition to being used for Migraine headaches. Some children with ASD may be on this medication. It is also used in Bipolar disorder, as an alternative to Lithium.

Taking magnesium and Mefennamic Acid (Ponstel) together may increase the amount of Mefennamic Acid absorbed, possibly leading to an increase in side-effects. Mefenamic Acid is a NSAID used for pain and PMS.

References

1   Karppanen H, Karppanen P, Mervaala E. Why and how to implement sodium, potassium, calcium, and magnesium changes in food items and diets? Institute of Biomedicine, Pharmacology, University of Helsinki. *J Hum Hypertens.* 2005 Dec;19 Suppl 3:S10-9.

2   Sloan Kettering Health Care Information for Professionals, See: www.mskcc. org/mskcc/html/11571.cfm?RecordID=481&tab=HC

3   See: www.umm.edu/altmed/ConsSupplements/Interactions/Magnesiumcs.html

24

|||||||||||||||||||||||||||||||||||||||||||||||||||||||||||||||||||||||||||

Testimonials

EDITOR'S NOTE: While there is considerable science behind transdermal magnesium therapy, it would be remiss of us to discount the accounts of people who have already experienced demonstrable shifts in the quality of their life after adding magnesium to their nutritional regimen. We hope that you are encouraged to realize your own success, by reading the successes freely shared by others.

The Premature Pronouncement

ISSUES: HEART CONDITION, DIABETESE, INFLAMMATION, CHRONIC PAIN

I am an older woman who has been diagnosed with a terminal heart condition, diabetes, high blood pressure, and neuropathy. Over a year ago, I had a major heart attack, which killed the entire back of my heart, leaving only a damaged front of my heart to carry the load.

My doctor told me that I would die shortly and to get my papers in order. His response when I disagreed with his prognosis was to offer more drugs to help me deal with my "denial," in essence taking away any hope I had to survive.

At that point, I decided to look for a cardiologist who would be open to alternative methods of healing, allow me to be involved, and would most importantly allow me to have hope. After months of unsuccessful appointments with closed-minded doctors, I decided to take my healing into my own hands, and began my quest for a form of natural healing on my own.

I stopped taking all prescription drugs and began experimenting with various alternative methods, techniques, and products, in hope of finding a "Miracle Cure". I felt no effect from some of things I tried, with others I felt a little improvement, but nothing made any real difference in my pain level or my incredible weakness. No matter what I did, I could not regain my strength or alleviate my continuous pain.

Then a friend of mine (bless his heart) sent me some magnesium oil. The first day I sprayed a little below my ears, on some glands that had been swollen for 20 years. Now I am serious, for 20 years, every single day, these glands had been swollen and sensitive to touch. When I got up the next morning, I decided to spray a little more on the glands, but when I began to rub the oil in, I noticed there was NO swelling. The swelling wasn't reduced after 20 years — it was suddenly gone!

That was the beginning of my love affair with magnesium oil. After two weeks of spraying it on my body 3-5 times a day, I now have periods each day where I am pain free, and I am discovering more strength each day! It is incredible! This is the first product I have found that has given me such immediate and powerful results.

I absolutely love this product and believe it has saved my life. I do not type these words casually; I was a dying woman and could feel my life force ebbing away each day. I now feel my strength and life force building in me, and know that I will heal.

I believe this magic oil knows no limits in healing, and I wish the blessing of magnesium oil on anyone who is sick or in need of healing.

When we first started talking about the magnesium I was dying. I knew it inside. I am no longer dying. I feel life in me. I am so happy.

J. Jones, Washington State

Uncramping His Style

ISSUES: CRAMPS, MUSCLE ACHES AND PAIN

Interestingly, right when the magnesium oil came I was having a strong cramp in my neck and shoulder, which I sometimes get from the computer. When I get it, it usually comes on for a day, until it is excruciating, and nothing makes it go away except the passage of a couple days' time. Well, I thought, I'll see if this magnesium stuff works like they say, and I rubbed one small squirt into my left neck and shoulder. Within 5 minutes the pain was gone, and I did not get several days of excruciating pain, like I usually do.

Since that time I've used it on a couple other muscle aches with success, a couple of skin scratches that weren't healing very fast, and they were healed in a couple of days. I've tried it on a nagging joint pain in my left shoulder, which it hasn't helped so far. I did use it last night on a sore throat I felt coming on, and it was gone before I went to bed. I think you may be onto something with this!

Skip J.

Soothing Nights

ISSUES: INSOMNIA

We have just started using the magnesium products. The bath is especially relaxing. Elaine and I sleep the night through after an evening 20-minute soak in 2 oz. magnesium bath salts in hot bath. Elaine usually gets up often in the night. I am also spraying magnesium oil onto my toothbrush with toothpaste, as well as in mouthwash. I also use it in my niddy pod, a little tea-pot looking container filled with water and some sea salt, the pours through one nostril and exits the other. Elaine is using the gel on her feet to relieve the peripheral neuropathy pain and hopefully rebuild the nerve cells.

Ken Norton

One Step at a Time

ISSUES: AUTISM

I've just started using the magnesium oil on my 7yr old ASD (Autism Spectrum Disorder) son. He's always tested very low in magnesium and I don't believe oral supplementation is doing that much. I put a few tablespoons of the oil in his bath water, and I also spray it onto my hands & rub it into his skin (tops of his feet & elbows). The reason I chose his elbows was because he's had this rash (large, bumpy, flesh colored) for quite some time. The magnesium stung at first when I rubbed it on, but after just a few nights, the rash is gone from one elbow and fading from the other!

Rose Langford

Deeper Muscle Relaxation Yields Better Vision

ISSUES: VISION ACUITY

The oblique muscle actually loops around the eye through a loop of tissue under the forehead. Its purpose is obviously not to orient the eye as in training on an object, but to actually squeeze the eye like a belt around a water balloon. This gentle squeezing produces a tiny, less than a millimeter change in the length of the eye and actually lengthens it for near focusing. A nearsighted person has chronic partial spasms in this muscle so it never completely releases. Therefore the eye is always configured for near vision. Nearsighted people often have pain above the eye under the eyebrow due to this tension and stress. Conversely, a farsighted person sometimes experiences pain in the temples where some of the oblique muscles underlie. Hypertonic oblique muscles reflexively inhibit the oblique muscle and the eye is always predisposed to distant vision and eventually due to lack of innervations, the oblique muscle becomes more and more useless.

After reading about unwinding spasms with magnesium oil, I diluted a little in an eyedropper bottle until the salinity was neutral to my tears and comfortably dropped it into my eyes. The effect is subtle, but I experienced some of my best distant vision (my challenge is nearsightedness) a day after using the drops for two days.

Sam Patterson

Experience Speaks for Itself

ISSUE: PARKINSON'S DISEASE

I have completed my first day of magnesium oil therapy on William who has had Parkinson's for over 20 years. I am hoping for a revival of functionality but not with high expectations because of the severity and duration of his symptoms.

His condition before starting the magnesium oil was: He couldn't talk at all. Could not articulate whatsoever! He was barely functional and did nothing voluntarily. No exercise and no attempt to stop drooling. The drooling was getting so bad and so constant that I was beginning to isolate him to his bedroom in his big recliner because the carpets are new here and the enzymes of the saliva stain permanently. It appeared to be getting worse by the week. That's how he was. He also had started getting violent with me. If I pushed him too hard he would fly into a rage and hit me with whatever he could lay his hands on.

I applied the magnesium oil twice yesterday and he woke this morning and washed his own face, cleaned his teeth and put on his robe by himself — without being told to do these things. This is unheard of and hasn't happened for *two years*. What is more, he is not drooling. The drooling has been massive and absolutely uncontrollable for about a year. His swallowing reflex is simply going. He has had his nutritional drink, his coffee, his brain formula, fresh veggies, and scrambled eggs and hasn't drooled once. So, my hope is high. This is the best I've seen (in him) for a very long time.

After only three days interestingly, his speech has been much better over all. I am applying it faithfully 3 times a day all over him. I will just keep up the application and let time do the explaining. I am very encouraged by the improvement in speech. I honestly did not expect to see any results. His eyes are brighter, his concentration is longer and better and his speech is much improved. By no means has he become a "toastmaster" but at least he can string 2 or 3 words together now and does not freeze up completely.

Most recently he has been quite violent. For example he thrashed me over the head with a plastic ladle one day so quickly that he got in 6 or 7 good thwacks before I could snatch it away from him. But since starting the magnesium oil, his demeanor has improved immensely. No more surly ugly looks, no more stubborn refusals to swallow or do something that I ask him to do. Great improvement and best of all he is now able to communicate so he can tell me what he wants and needs.

I am glad that I ordered a gallon of the magnesium oil. I figure that it may take a gallon to see any meaningful results. After 3+ weeks his speech is still much improved. It seems to be stable now. He couldn't give any lectures at Harvard, but he can make himself understood as to what he needs or wants. As I said, prior to the magnesium treatment he couldn't speak well enough to communicate anything.

Nancy English Vinal

A Ray of Hope

Issue: ADHD

I wanted to first thank you from the bottom of my heart and soul for the magnesium oil. I do see a change in my son Dane when we use it. The best change that I have seen is when Dane soaks in a tub before bed time he sleeps about 75% better. That is a good thing because he has always had a problem with sleep. He is medicated to help him sleep but even then, he does not rest well, talking in his sleep, flipping and flopping all night long, but after a soak that is reduced greatly.

I'm not saying that he is normal and completely calm, but any difference is good. He is a very high energy child, way too much energy for one kid. He tells me that he loves his magnesium. He insists that he feels calmer inside. I think you are on to something with this transdermal magnesium chloride lotion.

Think about all of the kids out there with ADHD and their parents who are willing to try anything to help calm their child. The only thing is that it needs to be used ever day consistently. It just doesn't work as well if you are not consistent. I guess his body can't hold on to the magnesium very long. In fact if he uses it before bed time he is good all night and by morning he is not as calm. But then I have been spraying him when he gets home from school.

Beth

Baths Bring Return *From* 'Forever'

ISSUES: AUTISM & VARICOSE VEINS

I've used (magnesium oil products) extensively in my healing practice and have had outstanding results. Previously I'd used and recommended Epsom salts, as they were cheap and easy for my clients to obtain. However, I found that Epsom salts were drying on people's skin, especially at the high concentrations that I've found to be most useful for healing and detoxification. We also tried a number of sea salts, such as Dead Sea salts, but there's something subtly different about magnesium that makes it work better.

I also use it extensively with my son. He had a lot of impulse control, focus, etc. issues, so we had him tested at the local university. Among other things, he was diagnosed with Asperger's Syndrome, which is a high functioning form of autism. During the evaluation, he was tested for a number of potential organic causes for his symptoms. His urine provocation test came back very high in several heavy metals. Apparently this is common in kids with autism spectrum disorders, one theory being that they are not as efficient at purging these toxins from their systems as "normal" kids.

We've been successfully using biomedical approaches to treat him, including far infrared (FIR) sauna sessions and supplements. One of the more helpful treatments has been consistent magnesium baths. DAN protocol suggests using Epsom salts (if I understand correctly it's because it is magnesium sulfate and these kids need the sulfates as well as the magnesium). But the frequent Epsom salt baths were very drying on the poor kid's skin. I switched over to magnesium chloride for his baths... My son loves those baths! He doesn't have to rinse after the baths, even though we're using quite strong concentrations of magnesium. And his skin is now soft and moist despite having several baths each week.

The combination of the FIR saunas and the magnesium baths have been very helpful in detoxifying his system, and have really calmed his behavior. His teachers, neighbors and other family members have all commented on how much better he can focus and track now.

I recently started to periodically mix some Epsom salts into the baths as well, just to make sure he's getting the benefits of the sulphates. Yet we mainly rely on the magnesium chloride.

Also, my father passed away recently, and my Mom (who is in her 80's) has moved in with me. She'd gotten some very unsightly and uncomfortable varicose veins. She's not felt well enough to be able to get in and out of the bathtub, but I've been able to give her foot soaks in the Master's Miracle soap/neutralizer/magnesium flakes almost every day. I've been spraying the magnesium oil on her feet and lower legs after the foot baths and rubbing it in as well. We were both pleasantly surprised to see her varicosities have shrunk considerably after a couple of weeks of this treatment.

L.H.

Many Areas, All Helped

Issues: Arthritis Inflammation, Sinus, Headaches, Oral Health

I will tell you all the ways I have been using the magnesium oil, and I seem to be adding to them all the time. I have been 'curing' a couple of things and have lots of things to use it on. In fact it seems every week now, I am using it for more things.

I have been using the magnesium oil successfully on the pain and inflammation of arthritis in my knees. I have been rubbing lots of the magnesium oil on and around, and under my knees, a couple of times a day. Also to get rid of both sinus headaches and also what I call "ME" headaches or headaches caused by Myalgic Encephalomyelitis.

I have been rubbing it under my eyebrows, on my temples, on my forehead and behind my ears and the edge of my skull around the sides and back for the sinus headaches and congestion (being careful not to get it into the eyes though) plus the top of my spine and into the back of my head for the ME headaches.

Also I broke out in boils a few weeks ago and have been rubbing it on them. The boils have been very painful the first few days as the tops come off and they seem to have what look like large pores all over them; but it does work.

Those baths are terrific! I am sleeping better, in spite of the boils, than I have in 10 years. By better, I mean I seem to sleep deeper and feel more rested, plus don't wake up as often at night and fall asleep easier.

My sinuses don't get nearly as congested at night now — they tried everything to stop my almost continuous sinus infections but nothing worked. I have been using a homeopathic mixture which does work; but I have to use it every night and after 4 months realized that it was costing me a lot of money and while I was sleeping better, it wasn't curing anything.

I rub it onto my gums — I have gingivitis. My gums don't bleed anymore. I get a lousy taste in my mouth sometimes, but spay it 3 or 4 times in my mouth and it goes away.

I also get swollen and painful glands in my neck — didn't know I had so many glands until I got sick. Anyway, within 24 to 36 hours they will all clear up, after spraying the magnesium oil in the sides of my mouth and on my back molars, 3 times a day. I have gone for so long in the past with swollen glands that I have forgotten what my face really looks like. I also squirt it onto the back of my throat when I get a sore throat now too — it goes away. I was surprised — I tend to be skeptical. In my experience, things that work for one person don't necessarily work for others, but this magnesium chloride seems to be different.

Frequently the glands under my arms and behind my knees also get sore and swollen. I have been rubbing it on these glands and it has been causing the pain and swelling to go away. That keeps coming back but it comes back less and less and goes away quicker. I tried just rubbing the glands without the magnesium just to see if I could get the same results — but it didn't work.

When I first started using it on my face and neck it "burned" or stung a lot so I diluted it with filtered water by 50%; after a week or so, then started using the magnesium gel; now doesn't sting anymore even with the undiluted magnesium oil. My skin is much better; it does take away the wrinkles to a point. It sort of makes the skin smoother and makes the pores smaller. I have never found anything that truly does make the pores smaller before — those cosmetic face clay masks work for a few hours but my pores now seem to be permanently smaller.

It also exfoliates the skin. I know there are products out there in the cosmetic world that with much help from the consumer are supposed to do this — just never found one that worked nearly as well nor easily as the magnesium chloride oil. Can't remember when I last had such soft feet.

Shan Russell

Multifaceted Relief

ISSUE: DIABETES, HIGH BLOOD PRESSURE, MUSCLE CRAMPING, CHOLESTEROL

I have been using magnesium oil now for about three months, and find it keeps my elevated blood pressure much lower without all the pills the doctor wanted me to be taking. It acts as a natural calcium channel blocker as research is showing. I am a diabetic and find this very encouraging to prevent complications.

I spray all over the body every day, and take a full bath three times a week with magnesium oil added. I can tell the baths are strong, and when I get upset, it helps to calm me even more than baths did without the oil... when daily stresses get to me, I run for a relaxing bath with magnesium oil now.

My husband has a blockage of an artery in his leg, and often had trouble with pain and soreness. He is on his feet all day long, and he can feel the muscles in his leg cramping. He used to come home from work limping. He sprays the oil on his leg, and sometimes we massage his leg with this oil.... but even without the massage, he feels the difference within about 5 minutes of spraying the oil on. He says the oil itches at first, then the pain and cramping disappears. He has started to use the magnesium oil with a heat treatment, too, and lately he is not coming home limping anymore!

Research is also showing that magnesium has effects like statin drugs in lowering cholesterol. My husband's cholesterol used to be very high, and now it has also dropped significantly..... yes, he has improved his diet, but I can't help but wonder if magnesium oil is not also working its wonders for this! He refused the advice on using a prescription drug from his MD, and tried magnesium instead with a lowering of bad cholesterol by almost 250 points, and is now back to very near acceptable levels, a truly gratifying surprise and benefit of increasing magnesium levels. We are both thrilled beyond words.

Claudia French RN, LPHA

Extreme Athletics

ISSUE: MUSCLE ACHES, CRAMPS, AND PAIN, STRENGTH RECOVERY

I did the magnesium soak two days ago with 4 ounces. The next morning I was better. Yesterday I did a 101 mile bike ride up a 6200 foot mountain. Was out in the heat (90 degrees) for 7½ hours, and then ran two miles when I got home. I am usually cramped up when I get home after a day like this and feeling pretty done in, but no cramps. I did another soak after the ride and run (I used more as my gallon arrived), and today I am not sore at all. I should be limping around. I did an easy 2.5 mile ocean swim and took another bath with 6 ounces in the tub, then got a massage and my body is feeling good.

I am very suggestible, but I think this stuff works! It burns like hell when I put it on my chest and legs, but it stops after 15 minutes so I don't care. In the tub I get no (burning) reaction.

D. Minkoff

25

||||||||||||||||||||||||||||||||||||||||||||||||||||||||||||||||||||||||

Magnesium —
Conception to Death

I t is sad that we live in a medical world that ignores the evidence that magnesium is crucial for every single function in our own bodies. From the moment we are conceived to the moment we die, magnesium is at the heart of life as is water and air. It is that basic and that important so when we read that the majority of the pubic is deficient in magnesium we can only despair for doctors and public health care officials seem uninterested and are not likely to care. It is like the laws of gravity are being ignored and E does not equal MC^2 in the world of medical science and practice. It is simply incredible how intelligence and medical logic have been thrown out the window for blind belief systems that more than equal the kind that believed the world was flat.

How does magnesium impact each stage of development and bring us to either a longer life, or a more peaceful death? How do we accept this gift of life and drink of its benefits? These are questions we have endeavored to answer in this book because of the ever present need our bodies have for magnesium for proper functioning. Magnesium is disappearing from our food, and public health is being severely compromised because of this reality of modern life.

That this is being done deliberately is equivalent to any mortal sin; the sin of knowingly removing magnesium and other vital nutrients via food processing. It is equivalent to a race of aliens coming to our planet and sucking out the oxygen from the atmosphere. Worse, we find health officials acting to remove our access to vitamin and mineral supplements that are the ONLY way informed beings can compensate for the loss. They are targeting the vitamins and minerals as dangerous with totally exaggerated estimates of their toxicity instead of being honest about their use of poisonous medicines that are creating tidal waves of iatrogenic death and disease.

They might as well start turning off peoples' water taps because people can drown in too much water. For years there has been evidence of the increasing need for magnesium, yet the medical system goes on like a killer juggernaut not only ignoring the desperate need of the public to remineralize, but going as far as to cutting off access to mineral supplements.

Through the years the best information about magnesium has come from renowned magnesium researchers Dr. Mildred Seelig, and Dr. Jean Durlach. Seelig has watched, observed and researched every phase of life affected by magnesium. In this book we have written about magnesium's effect on birth, life and aging, sexuality, menopause, osteoporosis, various illnesses, such as cancer, diabetes and heart disease prevention. We have seen how

easily many diseases can be cured or avoided when we bring sufficient attention to our magnesium needs.

The lifesaving grace is that magnesium chloride is
available for all from the sea. Truly a gift of life!

Deficiencies of magnesium are known in our diets and are
seen in our bodies from a very young age. In this closing chapter
we will cover a wide range of medical and health situations where
magnesium is a factor. This could be an endless book simply because magnesium is so basic to life that its deficiency can be part
of the etiology of any, and every disease. Doctors are not trained to
think about medical basics, they do not monitor the water people
intake, nor the magnesium needs of the body.

One of the main messages of this book is that magnesium
is right up there in importance with air and water. With air we
only have minutes of leeway before we die from its deficiency.
With water we have perhaps a week. Magnesium deficiencies are
much slower to manifest but the devastation does come and results quickly, first in weakened resistance to diseases and then over
the long term descent into serious chronic diseases of many types.
This chapter addresses some of the syndromes left out from earlier
chapters and updates the materials on diabetes.

Pregnancy, Childbirth, and Childhood

Magnesium is needed for reproductive fertility[1,2] and the use
of pharmaceutical contraceptives is known to diminish magnesium
stores in our body.[3] The rate of premature births has increased more
than 30% since 1981,[4] but a central obvious cause is ignored by

doctors. Instead, infertility treatments, including the use of super-ovulation drugs and multiple embryos are more frequently used by mainstream medicine today, not magnesium. Magnesium plays a crucial role in fertility, pregnancy,[5,6,7] and in early newborn life[8] and many of the problems could be easily and simply resolved by magnesium supplementation.

In 1991 Dr. Jean Durlach said, "Primary magnesium deficiency may occur in fertile women. Gestational magnesium deficiency is able to induce maternal, fetal, and pediatric consequences which might last throughout life. Experimental studies of gestational magnesium deficiency show that [magnesium deficiency] during pregnancy may have marked effects on the processes of parturition and of post-uterine involution.

"It may interfere with fetal growth and development from teratogenic effects to morbidity: i.e. hematological effects and disturbances in temperature regulation. Clinical studies on the consequences of maternal primary magnesium deficiency in women have been insufficiently investigated."

Magnesium is frequently used as the treatment for stopping premature labor, and the seizures of eclampsia, at the point it starts, but might be more helpful in preventing these if supplemented throughout the course of pregnancy. Dr. Durlach has also shown the increased safety of using magnesium chloride over magnesium sulfate.[9] Even worse is the evidence that magnesium deficiency/depletion is involved in the etiology of Sudden Infant Death Syndrome (SIDS).[10,11,12,13]

*The evidence is clear that inadequate magnesium
intake is common during pregnancy and that
the plasma levels of magnesium tend to fall,
especially during the first and third
trimesters of pregnancy.*

Dr. Mildred S. Seelig

We really have to wonder when we find out that most prenatal vitamins don't even contain magnesium, and when they do its in pathetically small amounts. (25-50 mg.) It is so sad that despite all the evidence and the absolute ease with which the public could have their magnesium needs met, we do not even help the babies and expectant mothers. When we do find magnesium used correctly, i.e., when magnesium sulfate is routinely given intravenously to halt pre-term labor or prevent convulsions in women with pregnancy-induced hypertension (eclampsia) we see clear results. In 1987, Dutch researchers found magnesium prevents hemorrhaging in the brains of infants whose mothers have this form of hypertension. Several randomized controlled trials (RCTs) have provided compelling evidence that $MgSO_4$ is the drug of choice for maternal seizure prophylaxis in pre-eclampsia, whether preterm or term.[14]

Scientists have shown that giving magnesium sulfate to pregnant women may greatly reduce the incidence of cerebral palsy in infants born weighing less than 3.3 pounds. These low birth weight infants are 60 to 75 times more likely to develop cerebral palsy than babies that reach a normal weight before birth — and the number of children with cerebral palsy is growing. In studies done another surprising outcome was the reduction of mental retardation when magnesium was provided during pregnancy.[15]

Allergies and Asthma

Increasingly we see evidence of the effects of magnesium deficiency in childhood, which collide with other factors like constant exposure to toxic chemicals and the violent chemical invasions offered by vaccines. Dr. Allan Becker at the University of Manitoba looked at 14,000 boys and girls born in 1995 and found that as many as 14% had asthma. "We're talking about one in seven children — that's a huge proportion of the pediatric population," Becker says. "It's in every classroom, every school, and many, many families. It's huge." This reality is reflected in the population at large. Asthma affects about three million people in this country, six out of 10 of whom do not have control of their disease, according to the Asthma Society of Canada. It kills 500 people in Canada each year, 5,000 in the U.S. Meanwhile, the World Health Organization says 150 million people around the world have asthma, and over 180,000 die annually as a result of it.

The relationship between asthma and allergies is a strong one. Asthma is often triggered by allergies of all kinds, food allergies, pollutants, dust, mold, chemicals and pharmaceutical drugs. Magnesium losses are notorious for occurring as a result of the drugs used specifically for asthma that open airways and reduce inflammation. Magnesium is known to help relieve bronchospasm, or constricted airways in the lungs and has been used intravenously to help relieve the symptoms of life-threatening, drug-resistant asthma attacks and to diminish the effects of asthma drugs used.[16,17]

The global incidence of allergic diseases such as food intolerances, asthma, eczema and hay fever is going through the roof in comparatively well-to-do Western cultures, says Mark Jackson, director of the University of Exeter's Centre for Medical History in England, and author of the forthcoming book *Allergy: The History of a Modern Malady*. Something is leaving many of us with immune systems primed to overreact when confronted with a wide range of substances from mold, to bee pollen and dust. "If you go back to the beginning of the 20th century, the notion of allergy wasn't even around," Jackson says. "A hundred years later, we now think that between 20 and 30% of the Western population is allergic to something -- and the figures suggest, certainly in Europe, that perhaps 50% of the population will have some kind of allergy by 2015."[18]

Research published in the *American Journal of Epidemiology* in 2002 shows that when the diets of 2,566 children ages 11-19 were studied, less than 14% of boys and 12% of girls had adequate intakes of magnesium and low magnesium intake was associated with lower measures of several lung functions (including lung capacity and airway flow). Based on the growing body of evidence, the researchers suggest that consideration should be given to public health interventions to increase magnesium intake.[19] **Magnesium deficiency definitely accentuates the allergic situation**," says Terry M. Phillips, D.Sc., Ph.D., director of the immunogenetics and immunochemistry laboratory at George Washington University Medical Center in Washington, D.C., and author of *Winning the War Within*.

In a study conducted at Brigham Young University in Provo, Utah, researchers found that laboratory animals severely deficient in magnesium had much higher blood levels of histamine when exposed to substances that trigger allergies than animals getting sufficient magnesium.[20] "The flow of calcium into and out of a cell helps regulate some cell function," explains Dr. Kay Franz, one the study's authors. "So it's possible that a magnesium deficiency changes the permeability of mast cell membranes, allowing calcium to more easily enter cells. When that happens, histamine is released." "In animals, magnesium deficiency causes the release of substances that can act on immune cells such as mast cells and basophils and make them hyperactive — more likely to release histamine," he says. Magnesium deficiency also causes other immune responses in the body that can lead to severe and sometimes fatal consequences.[21]

Cystic Fribrosis

In Cystic Fribrosis[22] there are many problems that might cause magnesium deficiencies, which could then in turn affect survivability. Cystic fibrosis (CF) is a hereditary disease of exocrine gland secretion characterized by the production of viscous mucus secretions. Clinically, CF is most commonly characterized by recurrent pulmonary infections, pancreatic insufficiency with maldigestion and malabsorption, and excessive losses of sweat electrolytes.

Good nutrition status in CF is crucial to improving a patient's ability to fight infection and promote survival. The nutrition status of adults with CF is commonly well below average, so that wasting and malnutrition are significant clinical issues. The reasons for malnutrition in CF can be broadly divided into three categories: (1) energy losses, (2) decreased dietary intake, and (3) increased nutrient requirements.[23] Aminoglycoside antibiotics, frequently used for the recurrent pulmonary infections in CF, lead to a loss of magnesium, and patients on these drugs may be at risk for severe magnesium deficiency.[24] Children with CF have significantly lower median erythrocyte concentrations of sodium, magnesium and zinc and a higher median concentration of calcium than both the healthy control children and the parents of the CF children.[25] All of these findings should lead our doctors to the conclusion that magnesium should be increased in CF, yet we find that only multivitamins are suggested with low magnesium content.

The Emerging Truth

Magnesium, Metabolic Syndrome, and Diabetes

The rising concern about metabolic syndrome[26] (synonymous with pre-diabetes) is another concern increasing today, and magnesium's role simply cannot be ignored in these disorders, but it is! There is no doubt today that diet alone cannot ensure adequate intake of magnesium to prevent the problems associated with the explosion in rates of metabolic syndrome in youth and adults, or the prevention and the treatment of diabetes.[27] Metabolic syndrome is identified as an extremely high risk factor in developing Type II diabetes and concerns about the findings in children and adults are mounting to alarming levels with every passing day. This incredible surge in the number of adolescents diagnosed with metabolic syndrome has health professionals fearing for the future of America's youth and scrambling to find causes and cures for the epidemic.

More than one in four American adults have metabolic syndrome, according to the latest government estimates[28], and recent reports to the American Diabetes Association at their 2006 symposium report that half of all Americans will develop a condition known as insulin resistance, a type of "pre-diabetes" that puts them at high risk of heart attacks.[29]

The March 27, 2006 issue of *Circulation* discussed the role magnesium might play in preventing this serious syndrome. A 15-year study looked at the magnesium intake of 4,600 people aged 18-30. They found that people with the highest intake of magnesium had a 31% lower risk for metabolic syndrome when compared with the group with the lowest magnesium intake.

Dr. R. Keith Campbell, a distinguished professor of pharmacy at Washington State University in Pullman and an expert on diabetes mellitus told a recent gathering of pharmacists that clinicians should add magnesium supplements to the treatment of metabolic syndrome along with any weight loss treatments or drugs used in treatment of diabetes itself.[30] He firmly instructed on the necessity of the addition of magnesium to any new drug treatment or dietary attempts at weight loss. "People with elevated blood glucose levels—one of the abnormalities that constitute metabolic syndrome—lose magnesium in their urine" Campbell said. On top of that, he said, many of the foods recommended for patients with diabetes mellitus lack magnesium.

"Somewhere between 40% and 80% of people with diabetes," he said, "have low body stores of this intracellular cation," which affects numerous enzyme systems in the body, including those affecting blood pressure, lipids, and insulin resistance. A week after Campbell spoke; the American Heart Association journal *Circulation* published online the article suggesting an inverse relationship between magnesium intake by healthy young adults and the risk of metabolic syndrome in later years.

Hypertension

Epidemiologic studies confirm that hypertension
is correlated with low magnesium levels.

Through the years there has been extensive research on the effects of magnesium on hypertension and its calcium channel blocking effects have been well known and accepted. But supplementation with magnesium continues to be ignored in favor of prescription drugs.[31] These drugs treat the symptoms of hypertension and do not address the underlying cause, and even worse, often cause more depletion of magnesium themselves.

Fifty million Americans have high blood pressure. "Americans now 55 or over face a 90% chance of developing high blood pressure, or hypertension, a major risk factor for heart attacks, strokes, congestive heart failure, circulatory failure, kidney disease, and loss of vision." More than 15 million (more than 30% of the 50 million) are undiagnosed.[32]

Dr. Jay Cohen, author of *"Toxic Dose"* tells us in a recent Life Extension article, "High blood pressure is an even stronger predictor of cardiovascular risk than high cholesterol. Scientific studies directly correlate high blood pressure with decreased longevity. Yet most mainstream physicians and their patients ignore this risk until life threatening hypertension has already developed, though damage may be well established prior to discovery. He goes on to say that "a well-established body of research indicates that nutrients such as magnesium are highly effective in treating and—even more importantly—preventing high blood pressure". Because this metallic element is not plentiful in foods, magnesium supplementation may be effective in both preventing and controlling high blood pressure."[33] Others, such as Dr. Sherry Rogers and Dr. Carolyn Dean have written extensively about magnesium's benefit in controlling hypertension.

Dr. Lawrence Resnick, professor of medicine and director of hypertension, Wayne State University, studied the blood pressure of patients who were both diabetic and non-diabetic. He found that all patients with hypertension, whether diabetic or non-diabetic, had lower magnesium levels than people with normal blood pressure.

Resnick says he has treated patients who were hypertensive in spite of taking one or two medications to treat this disease. And that by adding magnesium their pressure returned to normal.[34] In addition he maintains that ionic disturbance might be the missing link responsible for the frequent clinical coexistence of hypertension, atherosclerosis and metabolic disorders. Ageing cells may become more vulnerable to ion disturbances, leading to possible elevation of intracellular free calcium and concurrent magnesium depletion. The "ionic hypothesis" of ageing supposes that an alteration in the cellular mechanisms which maintain the homeostasis of cytosolic calcium concentrations plays a key role in the ageing process, and that a sustained accumulation of cellular calcium and/or the depletion of cellular magnesium may also provide the final common pathway for many ageing-associated diseases, including hypertension and NIDDM.[35,36] Researchers at Case Western Reserve University report in the August 2002 issue of *Nature* how magnesium activates microscopic ion channels in the membrane of a cell. These particular ion channels are important in controlling blood pressure.

"Research of this kind may help us to understand why some therapies such as magnesium supplements are important in the prevention and management of hypertension or heart failure," said Jianmin Cui, the lead researcher and assistant professor in the department of biomedical engineering at CWRU. "Along with some other groups, we have discovered that when magnesium is applied to calcium-activated potassium channels, these channels will open. We know from literature that the opening of these channels can reduce blood pressure."

Hypertension, Cui explained, results from the contraction of blood vessels, which causes an increase in blood pressure. "The diameter of blood vessels is controlled by smooth muscle cells around them," he said. "When magnesium reaches these potassium channels, the channels open causing blood vessels to dilate and therefore reduce hypertension."[37]

Dr. Mildred Seelig concluded: "The studies that employed the larger supplements of magnesium did in fact show that it has ability to lower high blood pressure"[38] Magnesium prevents blood clotting and arterial spasm. Drug therapy in elderly patients with hypertension has become a universal recommendation. Lifestyle modifications, such as reduced salt intake, regular exercise, and controlling body weight, may not be effective in reducing complications of hypertension in elderly adults. Magnesium is even being added to salt preparations, with significant effects on lowering blood pressure.[39]

Doctors are not only continuing to ignore the importance of magnesium but they are actually making the situation worse with many of the drugs they routinely prescribe. "Few doctors know that diuretics help flush magnesium as well as potassium from the body. The resulting magnesium deficiency hinders potassium use by the cells. Magnesium deficiency keeps people from replenishing potassium," says hypertension expert Dr. Chris Mende.

Death and Dying – End of Life Issues

Magnesium deficit may participate in the
clinical pattern of aging - neuromuscular,
cardiovascular and renal symptomatologies.

In general, as we age our tendency to compound accumulative magnesium deficiency only rises, leaving us increasingly vulnerable to a wide range of disorders and in the final analysis, to a miserable death. Cellular senescence is the phenomenon where cells lose their ability to divide, in response to DNA damage. Cells will either senesce or self-destruct (apoptosis) if the damage cannot be repaired. Organismal senescence is the aging of whole organisms, and is generally characterized by the declining ability to respond to stress, increasing homeostatic imbalance and increased risk of disease. Because of this, death is the ultimate consequence of aging. All physiological processes deteriorate in the face of magnesium deficiency vastly accelerating the aging process and hastening the onset of death.

Magnesium treatment has been repeatedly
shown to reduce the incidence of both
temporary and permanent noise-induced
hearing loss[40] and supplementation is now
being found to significantly improve
acute onset hearing loss.[41]

Dr. Mildred Seelig postulated that magnesium deficiency early in life gives rise to chronic abnormalities that persist throughout life, increasing morbidity and mortality and shortening life (Seelig,

1977; 1977/1982; 1978; 1980). "Little attention has been paid to the special magnesium needs of old people, to whether magnesium inadequacy might contribute to the aging process, or to whether magnesium supplementation might have any beneficial effects in the aged".[42]

More recently Dr. David Killilea noted that few studies have investigated long-term magnesium deficiency in isolated cells throughout their entire lifespan. In his research he found that cells replicative capacity was decreased with magnesium deficiency and that the changes were not related to increased cell death (apoptosis) and that there was altered sensitivity to oxidative stress and changes in mitochondrial physiology. He concluded that magnesium deficiency alters cellular and mitochondrial function and accelerates the senescent phenotype and may promote or exacerbate age-related disease.[43] Others interested in the aging process, like Dr. Norman Shealy, have given us much information on the values of transdermal magnesium therapy because his experience is that transdermal application of magnesium chloride alone has the ability to raise an important marker of aging – DHEA levels.

It is widely researched and recognized that magnesium deficiency commonly occurs in critical illness and correlates with a higher mortality and worse clinical outcomes in the intensive care unit (ICU).[44]

Stress factors particularly likely to be encountered by the aged include chronic anxiety and worry, and the acute stress of bereavement. Regardless of the cause, stress increases catecholamine and corticoid release, which in turn cause magnesium loss.

Catecholamines also increase myocardial calcium uptake (Nayler, 1967). Since low magnesium/calcium ratios increase catecholamine secretion (Baker and Rink, 1975), a vicious cycle is thus established when magnesium deficiency pre-exists. Well accepted is the contributory role of stress to cardiovascular disease, including sudden unexpected cardiac death. Less well known is the role of magnesium loss in the damage caused by stress. Long-term suboptimal magnesium intake, to which adaptation had taken place, so that signs of deficiency that were present early but no longer existed, resulted in decreased tolerance of stress and shortened life expectancy. (Heroux et al, 1973).[45]

Some of the principle causes of magnesium deficiency in aging and critical illness are gastrointestinal and renal losses. As we age, our kidneys lose their efficiency at their regulation of magnesium and maintaining the necessary balance. Our ability to tolerate normal everyday stressors may decrease, we may lose our teeth[46] and ability to take in and digest food properly. When magnesium intake is curtailed or when there is intestinal magnesium malabsorption, the normal kidney reduces magnesium excretion to very low values in an attempt to conserve the precious mineral nutrient. But, when renal magnesium handling is impaired, hypomagnesaemia ensues because, unlike with calcium, equilibration with cellular stores does not occur for several weeks. In the human, magnesium absorption decreases with age. Around the age of seventy it becomes two-thirds of what it usually is at around the age of thirty. Exchangeable pools of magnesium are reduced in elderly patients. Iatrogenic secondary magnesium deficit is especially important as there is an over consumption of prescribed drugs among aged patients, many of which deplete magnesium. Through primary and secondary magnesium deficiency and depletion, aging correlates strongly with increasing deficits.

Pain from any number of conditions may interfere with appetite, exercise, and contribute to poorer intake of a proper diet. Magnesium can be helpful in relieving pain of all types, both chronic and acute. Magnesium acts as a noncompetitive antagonist of the N-methyl-D-aspartate receptor, which has been implicated in the transmission of pain, according to Dr. Vincent Crosby and colleagues at Nottingham City Hospital. The researchers gave either 500 mg or 1 g of a solution of magnesium sulfate intravenously to 12 cancer patients who were in pain despite the use of strong opioids and other drugs. Patients receiving 500 mg of magnesium experienced pain relief for up to 4 hours, with three patients reporting complete relief, two patients reporting partial relief, and only one patient reporting no relief. Patients receiving 1 g of magnesium also reported varying degrees of pain relief that lasted up to 4 hours.[47]

Understanding how magnesium helps to transmit nerve signals and plays a key role in relaxing muscles, it becomes easy to understand how pain can be eased with the use of transdermal magnesium therapy. Magnesium permits calcium to enter a nerve cell to allow electrical transmission along the nerves to and from the brain. Even our thoughts, via brain neurons, are dependent on magnesium. Calcium causes contraction in skeletal muscle fibers, and magnesium causes relaxation. When there is too much calcium and insufficient magnesium, you will get sustained muscle contraction, twitches, spasms, and possibly even convulsions.

Transdermal delivery of magnesium chloride is highly effective in pain relief, calming agitation, and is easier to use when oral intake of food may become impaired in old age or disease. It is much easier to apply magnesium oil on the skin of an elderly person than it is to submit them to force feeding of food, pills or intravenous administration of drugs to compensate for losses.

> *Many people needlessly suffer pain*
> *because they don't get enough magnesium.*
> Dr. Mildred Seelig

Dr. Saul Pilar, a general practitioner in Vancouver, says he too has seen magnesium lessen pain and spasms, reduce fatigue and improve sleep. "My aim is that 100% of my patients get enough magnesium either from food or supplements," he says. "I am optimistic that in the future more attention will be paid to this essential mineral." Compassionate intelligent medicine surely would put magnesium supplementation, including transdermal delivery, first in terms of all medications. The fact that it is last speaks miles and miles about the level of compassion and intelligence in modern allopathic medicine.

References

1   Folic acid, vitamins E, B$_6$ and B$_{12}$, iron, magnesium, zinc and selenium deficiencies have been known to cause infertility that is easily reversible with supplementation (McLeod, 1996).

2   Howard JM, Davies S, Hunnisett A. 1994. Red cell magnesium and glutathione peroxidase in infertile women- effects of oral supplementation with magnesium and selenium. *Magnesium Research* 7(1):49x57

3   Women are another group of drug consumers who should be especially concerned with drug-induced nutrient depletion. Few women know that oral contraceptives lower the levels of such vital nutrients as Vitamin B$_2$, B6, and B$_{12}$, Vitamin C, folic acid, magnesium and zinc. Mainstream hormone replacement (chiefly Premarin, but also Estratab and raloxifene) can also lead to deficiencies in Vitamin B$_6$, magnesium and zinc. Drugs That Deplete—Nutrients That Heal a review of the book *Drug Induced Nutrient Depletion Handbook* by Pelton et al. See: www.lef.org/magazine/mag2000/july2000_review.html.

4   Institute of Medicine advisory: Reducing preterm birth; Causes, Consequences and Prevention; National Academies Press, June 2006; See: www.nap.edu/catalog/11622.html

5   Rats kept severely magnesium depleted (receiving 1/200 the control magnesium intake) for the entire 21-day period of gestation had no living fetuses at term (Hurley and Cosens, 1970, 1971; Hurley, 1971; Hurley et al., 1976). The shorter the duration of the magnesium deficiency, the fewer implantation sites were affected. When the deficiency was maintained from day 6-12, about 30% of the implantation sites were involved and 14% of the full-term fetuses had gross congenital abnormalities (cleft lip, hydrocephalus, micrognathia or agnathia, clubbed feet, adactyly, syndactyly, or polydactyly, diaphragmatic hernia, and heart, lung, and urogenital anomalies). Milder magnesium deficiency (1/130 control intake) maintained throughout pregnancy resulted in resorption of half the implantation sites and malformation of the living young at term.

6   *Magnesium Deficiency In The Pathogenesis Of Disease*, Seelig, M; Part 1, chpt. 2. www.mgwater.com/Seelig/Magnesium-Deficiency-in-the-Pathogenesis-of-Disease/chapter3.shtml

7   There is mounting evidence of magnesium insufficiency during pregnancy. Experimental acute magnesium deficiency has caused increased parathyroid secretion and even parathyroid hyperplasia (Larvor et al., 1964a; Kukolj et al., 1965; Gitelman et al., 1965, 1968a,b; Lifshitz et al., 1967; Sherwood et al., 1970, 1972; Targovnik et al., 1971). Thus, the possibility that magnesium deficiency is contributory to hyperparathyroidism of pregnancy, which is common despite widespread supplementation with calcium and vitamin D.

8   Infants at greatest risk of neonatal hypomagnesemia are low-birth-weight infants, including those suffering from intrauterine growth retardation (IUGR) or premature infants recovering from birth hypoxia or later respiratory distress, and infants born to very young primiparous women or to young mothers who have had frequent pregnancies or multiple births, to preeclamptic mothers, and to diabetic mothers. The incidence of neonatal magnesium insufficiency may be greater than suspected. The tendency of women with preeclampsia or eclampsia to develop

rising plasma magnesium levels during the last month of pregnancy, even without magnesium therapy, despite which they retain high percentages of parenterally administered pharmacologic doses of magnesium, suggests that magnesium deficiency might be far more common during pregnancy than is indicated by the incidence of hypomagnesemia. *Magnesium Deficiency In The Pathogenesis Of Disease*, Seelig, M; Part 1, chpt. 2. http://www.mgwater.com/Seelig/Magnesium-Deficiency-in-the-Pathogenesis-of-Disease/chapter3.shtml

9   Magnesium chloride or magnesium sulfate: A genuine question ; *Magnesium Research*. Volume 18, Number 3, 187-92, September 2005

10   Sudden infant death syndrome (SIDS) is defined as the sudden death of an infant or young child, which is unexpected by history, and in which a thorough postmortem fails to demonstrate an adequate cause for death. SIDS accounts for 35% of post-natal deaths.

11   Two clinical forms of chronic gestational Mg deficiency in women have been stressed: Premature labor when chronic maternal Mg deficiency is involved in uterine hyper-excitability, Sudden Infant Death Syndrome (SIDS) when it is caused by either simple Mg deficiency or various forms of Mg depletion. If gestational Mg deficiency is the only cause for uterine overactivity, nutritional Mg supplementation constitutes the etiopathogenic atoxic tocolytic treatment. Mg deficiency or various forms of Mg depletion. SIDS may be caused by the fetal consequences of maternal Mg deficiency through an impaired control of Brown Adipose Tissue (BAT) thermoregulation, mechanisms leading to a modified temperature set point. SIDS may result from dysthermias: hypo- or hyperthermic forms. A possible prevention could rest on simple maternal nutritional Mg supplementation. SIDS might be linked to an impaired maturation of both the photoneuroendocrine system and BAT. A preventive treatment of this form of SIDS should associate atoxic nutritional Mg therapy for pregnant women with total light deprivation at night for the infant. New data on the importance of gestational Mg deficiency. Durlach; *Magnes Res* 2004 Jun;17(2):116-25 EntrezPubMed.

12   Published findings in mothers of victims of sudden infant death syndrome (SIDS) and in the SIDS victims are compared with characteristics of magnesium deficiency in humans and animals. Observations concerning the level of magnesium in traditional diets of selected ethnic groups with the highest or lowest rates of SIDS appear to confirm the importance of magnesium in protecting the offspring from sudden death. The apparent impact of gestational magnesium (Mg) deficiency on the sudden infant death syndrome (SIDS);Cardell; *Magnes Res* 2001Dec;14(4):291-303

13   Magnesium deficiency promotes muscle weakness, contributing to the risk of sudden infant death (SIDS) in infants sleeping prone. Cardell; *Magnes Res* 2001 Mar;14(1-2):39-50

14   Magnesium Sulfate in Obstetrics: current data *J Gynecol Obstet Biol Reprod* (Paris). 2004 Oct;33(6 Pt 1):510 7

15   See: www.sciencenews.org/pages/sn_arch/12_14_96/fob1.htm

16   Oral magnesium supplementation helped to reduce bronchial reactivity to methacholine, to diminish their allergen-induced skin responses and to provide better symptom control in pediatric patients with moderate persistent asthma treated with inhaled fluticasone. *European Journal of Clinical Nutrition* advance online publication,

21 June 2006; doi:10.1038/sj.ejcn.1602475. Entrez Pubmed

17   Studies clearly show the efficacy of intravenous administration of magnesium, and use of inhaled (nebulized) magnesium, either added to other inhaled asthma drugs, or independently, are being considered. More data is needed. Cochrane Database Syst Rev. 2005;(3):CD003898. Inhaled magnesium sulfate in the treatment of acute asthma. Entrez Pubmed

18   Almost 50% of infants today suffer from some form of eczema, and the prevalence of hay fever stands at between 30% and 40% of the population — a two- to threefold increase in the last few decades. (In 2003, 18.4 million American adults were diagnosed with hay fever, as were 6.7 million children in 2004.) Health Canada estimates that non-food allergies are "the most common chronic condition in Canadians 12 years of age and older." The Allergy, Genes and Environment Network, or AllerGen, is Canada's response to the crisis. Part of the country's Networks of Centres of Excellence, Hamilton-based AllerGen is comprised of more than 100 scientists at 20 universities and research facilities across the country. Dr. Judah Denburg is AllerGen's scientific director. *The Allergy Epidemic*; June, 2006; See: www.macleans.ca/topstories/ health/article.jsp?content=20060605_128132_128132

19   Gilliland, F.D. et al. Dietary magnesium, potassium, sodium and children's lung function. *American Journal of Epidemiology*. 2002; 155: 125-131.

20   A synergism of antigen challenge and severe magnesium deficiency on blood and urinary histamine levels in rats.Wei W, Franz KB. *J Am Coll Nutr.* 1990 Dec;9(6):616-22. Entrez Pubmed

21   *Crit Care Med.* 1995 Jan;23(1):108-18. Entrez Pubmed

22   Cystic fibrosis (CF) is the most common fatal genetic disease in white populations (1 in 2,500 live births). CF occurs due to recessive mutations of the cystic fibrosis transmembrane conductance regulator (Cftr) gene, which encodes a transmembrane chloride channel expressed in the epithelium of multiple organs (1). The role of Cftr in the pathogenic process is unclear. Classic CF lung disease presents with a progression of inflammation and infection, and a decline in lung function marked by mucopurulent plugging, bronchiectasis, and intermittent bronchopneumonia. Recurrent exacerbation of lung infection and respiratory failure are the major causes of hospitalization and death. Improvements in antibiotic therapy, physiotherapy, and nutrition have enhanced the duration and quality of life of patients with CF. Effective management of airway inflammation could augment current therapy. Lung Inflammation as a Therapeutic Target in Cystic Fibrosis *American Journal of Respiratory Cell and Molecular Biology*. Vol. 31, pp. 377-381, 2004   See: http://ajrcmb.atsjournals.org/cgi/content/ full/31/4/377.

23   Nutrition in the management of cystic fibrosis; *Nutrition Reviews*, Jan 1996 by Dowsett, Julie. See: www.findarticles.com/p/articles/mi_qa3624/is_199601/ai/ n8735342/pg_2

24   Nephrol Dial Transplant (2000) 15: 822-826 See: http://ndt.oxfordjournals.org/ cgi/content/full/15/6/822#R9

25   Low concentrations of sodium and magnesium in erythrocytes from cystic fibrosis heterozygotes. Foucard, et al; *Acta Paediatr Scand.* 1991 Jan;80(1):57-61

26   Metabolic syndrome is a cluster of symptoms including large waist

circumference, high blood sugar, high blood pressure, low levels of HDL (good cholesterol) and high triglycerides. People with metabolic syndrome are at increased risk for heart disease and diabetes. Insulin resistance and abdominal obesity, according to the American Heart Association and the National Heart, Lung, and Blood Institute, seem to be the predominant risk factors for metabolic syndrome.

27   A 2005 study on Magnesium deficiency and insulin resistance in obese children states that hypomagnesemia (serum magnesium <0.78 mmol/l) was present in 27% of healthy lean children and 55% of obese children, indicating that serum magnesium deficiency may be more prevalent in children than previously suspected and definitely plays a role in insulin resistance. They conclude that serum magnesium deficiency in obese children may be secondary to decreased dietary magnesium intake. And more importantly that magnesium supplementation or increased intake of magnesium-rich foods may be an important tool in the prevention of type 2 diabetes in obese children. See: http://care.diabetesjournal.org/cgi/content/full/28/5/1175#T2

28   Prevalence of the Metabolic Syndrome Among U.S. Adults Earl S. Ford, MD, MPH, Wayne H. Giles, MD, MS and Ali H. Mokdad, PHD 2004; From the Division of Adult and Community Health, National Center for Chronic Disease Prevention and Health Promotion, Centers for Disease Control and Prevention, Atlanta, Georgia by the American Diabetes Association *Diabetes Care* 27:2444-2449, 2004

29   Defining A Crisis: Cardiometabolic Risk How Insulin Resistance Threatens Half Of Americans See: cardiometabolic.06.12.pdf

30   Treatments for Metabolic Syndrome May Expand. American society of Health System Pharmacists: May, 2006; See: www.ashp.org/news/ShowArticle.cfm?cfid=1656 649&CFToken=67389440&id=15117

31   There are more than 64 million annual prescriptions for calcium channel blocking drugs such as (Procardia, Cardizem, Norvasc, Verpamil, Adalat, Dilacor, Verelan, Calan), with sales exceeding $2.5 billion. [*American Druggist* 1997]

32   Lists numerous reliable sources of statistical information on hypertension: See: www.wrongdiagnosis.com/h/hypertension/prevalence.htm

33   Magnesium in Hypertension Prevention and Control; Jay S. Cohen, MD; *LE Magazine* September 2004. See: www.lef.org/magazine/mag2004/sep2004_report_ magnesium_01.htm

34   Magnesium fights range of serious ills; Dr. W. Gifford-Jones, Special to The StarPhoenix Saturday, April 29, 2006 See: www.canada.com/saskatoonstarphoenix/ news/weekend_extra/story.html?id=7e9bedf0-2513-4413-b34a-0881216c0cb6&p=2

35   *J Am Geriatr Soc.* 2000 Sep;48(9):1111-6. Cellular ionic alterations with age: relation to hypertension and diabetes. Entrez Pubmed

36   Electrolyte balance is a critical issue in managing comorbid conditions in both diseased and elderly patients. Patients with hypertension and diabetes need careful regulation of their calcium and magnesium levels, whereas in patients with congestive heart failure, sodium and potassium levels also are critical Potassium, magnesium, and electrolyte imbalance and complications in disease management. Weglicki W, Quamme G, Tucker K, Haigney M, Resnick L. Clin Exp *Hypertens*. 2005 Jan;27(1):95-112. Entrez Pubmed

37   Case Western Scientists Reveal How Magnesium Works On Ion Channels

Important For Regulating Blood Pressure; 2002-08-26 See: www.sciencedaily.com/releases/2002/08/020826071458.htm

38   Seelig MS, Rosanoff A. *The Magnesium Factor*. New York, NY: Avery Publishers; 2003.

39   Swapping plain old table salt for a blend of sodium chloride, potassium chloride, and magnesium sulfate can reduce systolic blood pressure by more than most dietary changes and as much as some antihypertensive drugs. Low cost and "This is actually a large effect, larger than is typically seen for dietary interventions in trials," Dr. Neal said. Salt Substitute Cuts Systolic Blood Pressure; *Medpage Today*: March 14, 2006; See: www.medpagetoday.com/tbprint2.cfm?tbid=2853

40   Magnesium provides significant protection against temporary threshold shift, complementing the previous permanent threshold shift human study. Both human noise-induced hearing loss studies introduced a novel, biological, natural agent for prevention and possible treatment of noise-induced cochlear damage in humans Reduction in noise-induced temporary threshold shift in humans following oral magnesium intake. *Clin Otolaryngol Allied Sci.* 2004 Dec;29(6):635-41 Entrez Pubmed

41   In a prospective, randomized, double-blind, placebo-controlled trial, 28 patients with idiopathic sudden sensorineural hearing loss were treated with either steroids and oral magnesium (study group) or steroids and a placebo (control group). Compared to the controls, the magnesium-treated group had a significantly higher proportion of patients with improved hearing (>10 dB hearing level) across all frequencies tested, and a significantly greater mean improvement in all frequencies. Analysis of the individual data confirmed that more patients treated with magnesium experienced hearing improvement, and at a larger magnitude, than control subjects. Nageris BI, Ulanovski D, Attias J. *Ann Otol Rhilol Laryngol* 2004;113:672-675.

42   Possible Role of Magnesium in Disorders of the Aged; Mildred S. Seelig, M.D Volume 3a, Modern Aging Research Intervention In the Aging Process, Part A: Quantitation, Epidemiology, and Clinical Research ; pages 279-305 available at: www.mgwater.com/aging.shtml

43   Magnesium Deficiency Accelerates Cellular Senescence In Human Fibroblasts See: www.americanaging.org/2005/kililea.pdf

44   Magnesium deficiency in critical illness. *J Intensive Care Med*. 2005 Jan-Feb;20(1):3-17. Entrez Pubmed

45   Role of Magnesium in Disorders of the Aged; Mildred S. Seelig, M.D., M.P.H., F.A.C.N. See: www.centerforantiaging.com/Role_of_Magnesium_in_Disorders_of_the_Aged.htm

46   The prevalence of edentulism, having no natural teeth, was higher for people age 85 and over (38 percent) than for people age 65-74 (24 percent). Socioeconomic differences are large. 46% of older people with family income below the poverty line reported no natural teeth compared with 27% of people above the poverty line.

47   *Journal of Pain and Symptom Management*, Jan. 2000; 19:35-39. Intravenous Magnesium Relieves Neuropathic Pain In Cancer Patients See: https://www.cancerpage.com/news/article.asp?id=328

Appendix

||||||||||||||||||||||||||||||||||||||||||||||||||||||||||||||||||||||||||||||||||||||

Magnesium Deficiency Questionnaire

With estimates that between 68% and 80% of the population is currently magnesium deficient, we have included the following questionnaire to help you determine where you stand.

Simply circle each "yes" answer on the list below that applies to you. Each is given a numerical value. When you finish, total your score.

--With 30-50, you likely have low magnesium.

--Over 50 & you most certainly have low magnesium.

YES	QUESTION...
2	Under excessive emotional stress
3	Irritable, or easily provoked to anger
2	Restless, or hyperactive
4	Easily startled by sounds or lights
2	Difficulty sleeping
3	Chronic headaches or migraines
2	Convulsions
3	Fine tremor or shakiness in your hands
3	Fine, barely noticeable muscle twitching around your eyes, facial muscles, or other muscles of your body
3	Muscle cramps
3	Muscle spasms in hands or feet
4	Gag or choke from spasms in your esophagus(food tube)

3	Have asthma or wheezing
2	Suffer from emphysema, chronic bronchitis, or shortness breath
5	Have osteoporosis
3	Have you ever had a kidney stone
2	Suffer from chronic kidney disease
4	Have diabetes
3	Have an overactive thyroid, or parathyroid gland
3	Have high blood pressure
4	Have mitral valve prolapse ("floppy heart valve")
3	Have very fast heart beats, irregular heart beats, or arrhythmia
3	Take Digitalis (Digoxin)
5	Take any kind of diuretic
5	Recent radiation therapy or exposure
4	Have more than 7 alcohol drinks weekly
3	Have you ever had a drinking problem?
2	Have more than 3 servings of caffeine daily
2	Eat sugar containing food daily
2	Crave carbohydrates &/or chocolate
2	Crave salt
2	Eat a high processed food/ junk food diet
2	Eat a diet low in green, leafy vegetables, seeds, & fresh fruit
2	Eat a low protein diet
2	Pass undigested food or fat in your stools
3	Suffer from chronic intestinal disease, ulcerative colitis, Crohn's, irritable bowel syndrome
3	Frequent diarrhea or constipation
3	Suffer from PMS or menstrual cramps
2	Pregnant or recently pregnant
4	In previous pregnancy had high blood pressure or pre-eclampsia
2	Chronic fatigue
2	Muscle weakness
2	Cold hands &/or feet
2	Numbness in face, hands, or feet
2	Persistent tingling in body
2	Chronic lack of interest, indifference, or apathy
2	Poor memory
2	Loss of concentration
3	Anxiety
2	Chronic depression for no apparent reason
2	Feelings of disorientation as to time or place
2	Feel your personality is stiff or mechanical
2	Hallucinations
2	Feel that people are trying to harm or persecute you
2	Face pale, puffy, or lacking in color
2	Loss of considerable sexual energy or vitality
2	Been told by your Dr that your blood calcium is low
3	Been told by your Dr that your blood potassium is low
2	Take Calcium supplements regularly without magnesium
2	Take iron or zinc supplements regularly without magnesium
2	Know chronic exposure to fluorides
3	Frequently use antibiotics, steroids, oral contraceptives, Indomethacin, Cisplatin, Amphotericin B, Cholestyramine, synthetic estrogens

<-- TOTAL ...

Subtract 15 from your score if you daily supplement at least 600 mg of magnesium

Magnesium Chloride Therapy

Raul Vergini, MD - Italy[1]

A French doctor, A. Neveu, observed that magnesium chloride has no direct effect on bacteria (i.e. it is not an antibiotic). Thus he thought that its action was a specific, immune-enhancer, so it could be useful, in the same manner, against viral diseases. So he began to treat some cases of poliomyelitis, and had good results.

Then he found the same good results in: pharyngitis, tonsillitis, hoarseness, common cold, influenza, asthma, bronchitis, broncho-pneumonia, pulmonary emphysema, "childhood diseases" (i.e., whooping-cough, measles, rubella, mumps, scarlet fever...), alimentary and professional poisonings, gastroenteritis, boils, abscesses, erysipelas, whitlow, septic pricks (wounds), puerperal fever and osteomyelitis.

But the indications for magnesium chloride therapy don't end here. In more recent years other physicians (and I among these) have verified many of Delbet's and Neveu's applications and have tried the therapy in other pathologies: acute asthmatic attack, shock, tetanus (for these the solution is administered by intravenous injection); herpes zoster, acute and chronic conjunctivitis, optic neuritis, rheumatic diseases, many allergic diseases, spring-asthenia and Chronic Fatigue Syndrome (even in cancer it can be a useful adjuvant).

The preceding lists of ailments are by no means exhaustive; maybe other illnesses can be treated with this therapy but, as this is a relatively young treatment, we are pioneers, and we need the help of all physicians of good will to establish the true possibilities of this wonderful therapy. From a practical standpoint, please remember that only magnesium CHLORIDE has this "cytophilactic" activity, and no other magnesium salt; probably it's a molecular, and not a merely ionic, matter.

The solution to be used is a 2.5% Magnesium Chloride hexahydrate ($MgCl_2$-$6H_2O$) solution (i.e.: 25 grams / 1 liter of water). Dosages are as follows:

```
        Adults and children over 5 years old
                                      125 cc
           4 year old children 100 cc
           3 year old children 80 cc
         1-2 year old children 60 cc
      Over 6 months old children 30 cc
     Under 6 months old children 15 cc
```

These doses must be administered BY MOUTH. The only contraindication to Magnesium Chloride Therapy is a severe renal insufficiency. As the magnesium chloride has a mild laxative effect, diarrhea sometimes appears on the first days of therapy, especially when high dosages (i.e. three doses a day) are taken; but this is not a reason to stop the therapy. The taste of the solution is not very good (it has a bitter-saltish flavor) so a little of fruit juice (grapefruit, orange, lemon) can be added to the solution, or it can be even used in the place of water to make the solution itself.

For CHRONIC diseases the standard treatment is one dose morning and evening for a long period (several months at least, but it can be continued for years). In ACUTE diseases the dose is administered every 6 hours (every 3 hours the first two doses if the case is serious); then space every 8 hours and then 12 hours as improvement goes on. After recovery it's better going on with a dose every 12 hours for some days.

As a PREVENTIVE measure, and as a magnesium supplement, one dose a day can be taken indefinitely. Magnesium chloride, even if it's an inorganic salt, is very well absorbed and a very good supplemental magnesium source.

For INTRAVENOUS injection, the formula is: Magnesium Chloride hexahydrate 25 grams Distilled Water 100 cc. Make injections of 10-20cc (very slowly, over 10-20 minutes) once or twice a day. Of course the solution must be sterilized. This therapy gives very good results also in Veterinary Medicine, at the appropriate dosages depending upon the size and kind of animals.

The Role of Magnesium in Fibromyalgia

By Mark London

Magnesium is important for people with fibromyalgia. Not only is our daily intake low, but we eat a diet which increases the demand for magnesium. And unfortunately, urinary magnesium loss can be increased by many factors, both physical and emotional. Magnesium loss increases in the presence of certain hormones. Stress can greatly increase magnesium loss. Thus the chances are almost 100% that a person with fibromyalgia has a magnesium deficiency since such people often have high levels of stress and a disrupted hormonal system. Magnesium utilization is also increased by the presence of estrogen, and this might explain why many women are diagnosed with fibromyalgia after menopause, when estrogen levels would decrease.

Additionally, the sleep disruption which occurs in fibromyalgia might also affect magnesium utilization, as sleep deprivation has been shown to cause lower magnesium levels. The reason lack of sleep causes a magnesium deficiency is probably due to the lower amounts of growth hormone secretion which occurs due to a sleep disturbance, especially the type that is found in people with fibromyalgia.

Low levels of ATP have commonly been found in people with fibromyalgia, and it is believed that this plays an important role in many of the fibromyalgia symptoms. Thus, a magnesium deficiency would definitely be a factor in worsening those symptoms. Magnesium is extremely necessary for proper ATP synthesis, because ATP is stored in the body as a combination of magnesium and ATP, which is known as MgATP. ATP requires magnesium in order to be stable. Without magnesium, ATP would easily break down into other components, ADP and inorganic phosphate.

The brain heavily relies ATP for many functions. In fact, 20% of total body ATP is located in the brain. Thus, low levels of ATP can diminish brain cognitive functions, a common problem in people with fibromyalgia.

Adequate magnesium is necessary for proper muscle functioning. Magnesium deficiency promotes excessive muscle tension, leading to muscle spasms, tics, restlessness, and twitches. This is due to an imbalance of the ratio of calcium to magnesium, as calcium controls contraction, while magnesium controls relaxation. Plus, in fibromyalgia, changes are seen in the muscles, such as "significantly lower than normal phosphocreatine and ATP levels" and "values for phosphorylation potential ... also were significantly reduced." All of these same changes are found also in magnesium deficiencies.

It's because of magnesium's ability to regulate nerve functions that other fibromyalgia symptoms occur when its levels are sub-optimal. Migraine headaches, mitral valve prolapse, and Raynaud's phenomenon, all problems commonly found in people with fibromyalgia, are also problems that have been associated with a magnesium deficiency.

Without enough magnesium, nerves fire too easily, even from minor stimuli. Noises will sound excessively loud, lights will seem too bright, emotional reactions will be exaggerated, and the brain will be too stimulated to sleep, all symptoms commonly found in fibromyalgia. If the oversensitivity to light and noise reminds you of someone suffering from a hangover, they are one and the same problem, as alcohol is known for decreasing magnesium levels, and magnesium supplementation has been found to relieve hangover symptoms.

Magnesium is thus involved in many functions in the body, and so it's no wonder that the chemical brain imbalances in fibromyalgia somehow seem connected to processes involving magnesium. It's because magnesium is involved in so many processes in the body, that a deficiency has a spiralling effect. Low magnesium levels cause metabolic functions to decrease, causing further stress on the body, reducing the body's ability to absorb and retain magnesium. A marginal deficiency could easily be transformed into a more significant problem. Any stressful event could trigger magnesium loss, so one could postulate that stressful events which trigger fibromyalgia are doing so by creating a high loss of magnesium. Perhaps people in a fibromyalgia flare could be helped by additional magnesium.

Unfortunately, magnesium deficiency is not easily detected, as serum levels do not reflect the levels of magnesium in tissues. This is the reason why it is so overlooked and ignored, both by doctors and by studies. And unfortunately, oral magnesium supplementation can be difficult because of absorption problems.

Digestion and diet play a key role in absorption. People with fibromyalgia often have conditions like Irritable Bowel Syndrome, gluten intolerance, or other problems that might limit absorption. Excess amounts of certain substances, such as fructose, may interfere with magnesium absorption. Phosphate can bind to magnesium in the gut, creating magnesium phosphate, an insoluble salt that can't be utilized. Many forms of oral magnesium supplements are hard to assimilate. The most common, magnesium oxide and citrate, happen to be the worst to assimilate, which is why both have a strong laxative effect. If you suffer from diarrhea from taking oral magnesium, it is often not because you are taking too much, but because you are not assimilating it well.

While most symptoms which are directly due to magnesium are reversible, magnesium indirectly causes problems that may not be reversible. Combined aluminum intoxication with calcium-magnesium deficiencies is not reversible through oral magnesium supplementation. Chronic magnesium deficiency can produce irreversible lesions in the brain.

I personally started taking magnesium for spasms and facial tics, only doing so on my own after neurologists simply told me to either get better sleep or take a prescription drug. The magnesium helped almost immediately, and I then slowly increased the dose to about 225% the RDA (balanced with 100% calcium RDA) At that point, all spasms and tics stopped completely, and they have not returned since starting that dose several years ago.

I doubt any traditional doctor would have been willing to prescribe that much magnesium. The RDA is 400mg, but many people believe this is too low. Traditionally, it's been recommended to take calcium and magnesium in a ratio of 2:1, because that is the ratio that these minerals are found in bone. But magnesium is less easily absorbed than calcium, so this ratio may not be valid for a lot of people, and in fact many calcium/magnesium combinations found in health food stores often have additional magnesium.

Magnesium is just one of many helpful remedies and/or supplements for that might be helpful for fibromyalgia. It's not a cure, but it may be helpful in relieving some of the symptoms.

Product Information

This book introduces a new magnesium product, a magnesium chloride taken from the ocean. It is very different from the crystal or powdered magnesium chloride created industrially with hydrochloric acid. The magnesium oil is considerably less toxic in terms of heavy metals than industrial fabricated magnesium chloride and is considerably more concentrated. There is approximately twice as much elemental magnesium as in magnesium oils that are created from magnesium chloride powder or crystal. Ocean-derived magnesium oil is between 31 to 35% magnesium chloride by weight.

A product that I have used personally that combines high dosage oral magnesium with natural detoxification and chelation is Chelorex™. Chelorex is safe and effective approach to heavy (toxic) metal chelation. When building a mega-magnesium protocol using Chelorex for oral magnesium and general detoxification is a good idea. Chelorex is an excellent source of magnesium, selenium and zinc, plus ascorbic acid, vitamin E and a whole list of natural agents and amino acids that facilitate detoxification of the body.

We are all exposed to toxic metals from a polluted environment everyday, from the air we breathe, the water we drink and other possible sources depending on where we live or work. For more information please visit:

www.imva.info/mercury.shtml

scienceformulas.com/

If you needed that extra nudge to start feeding your kids organic food, here it is: In a recent U.S. EPA-funded study, 23 Seattle-area youngsters were switched to an all-organic diet, and the levels of pesticides in their bodies declined to essentially zero after only five days. When the kids started eating conventionally grown food again, their pesticide levels shot back up. The study, published in Environmental Health Perspectives clearly shows that pesticide-free food leads to pesticide-free kids. United Press International, Christine Dell'Amore, 22 Feb 2006

I was lucky enough to be found by www.bulknuts4you.com who are a good source of foods high in magnesium. I seriously recommend the purchase of brazil nuts and toasted sesame seeds (very high in magnesium and calcium) and in general use organic whole grain foods, all of which you will find at this site.

For everyone who has not heard of nutritional yeast I highly recommend it as a source of B vitamins and many other wonderful nutritional things. It is up there with spirulina in my eyes. It is exceptionally good tasting and can be used in or on almost everything.

References

1   See: www.navi.net/~rsc/mgcl2_txt.html

Index

Symbols

γ-glutamyl cysteine 98
γ-glutamyl transpeptidase 98

A

Abilify 181
abscesses 345
absorbability 203
absorption 34, 350
absorption, transdermal 253
acetylcholine 3
Achilles tendon 254
acidosis, metabolic 28, 133
ACTH 265
acupuncture 169
Addison disease 300
Adenosine triphosphate (ATP) 4, 89, 127, 167
ADHD 174, 312
adolescence 54
ADP 348
adrenal cortex 262, 269
adrenal gland 237-238
adrenaline 263
adrenals 240, 263
adrenals, weak 245
Adrenocorticotrophic hormone 265
age reversal 242
aggression 174
aging 191, 233, 235, 238
agranulocytosis 165
agribusiness 23
Agriculture, U.S. Department of 77
AIDS xvi
ALA 110–111, 115
albumin 133
alcohol 25, 27
alcohol decreases magnesium levels 349

Aldosterone 237, 262
algae cell walls 114
Alka-Seltzer Plus Cold Medicine 287
allopathic 66
allopathic medicine 26, 39, 66–67, 75, 83
allopathic philosophy 26
Alloxan 140
alpha-lipoic acid 110
alternative medicine 76
alternative practitioners 75
Altura, Dr. Burton M. 15, 284
aluminum 129, 350
Alzheimer's disease 4, 100, 238
American Academy of Pediatrics 54
American Board of Clinical Metal Toxicology xv
American Cancer Society 55
American Diabetes Association 59
amino acids 5, 193
amino acids, brain 194
amnesia 173
amputations 144
amputations, lower-limb 145
anaphylactic reactions 43
angina 10, 65, 73, 148
angina pectoris 164
angiocholitis 43
Anhui Medical University 91
anorexia 176
antacids 301
anti-hypertensives 166
anti-platelet 166
anti-psychotic sales 173
antiarrhythmic 14
antibiotics 15
antidepressant (effects of magnesium) 100
antidepressant sales 173
antigen 230

antimony 255
antioxidant 68, 95
anus 244
anxiety 4, 100
anxiolytic 100
apathy 173
*A Physician's Guide to Natural Health
    Products That Work* 87
apprehension 174
Archives of Internal Medicine 260
Arizona, University of 84, 87
arrhythmia 5, 71, 74, 151, 166, 179,
    255
arrhythmia, cardiac 65
arsenic 139
arsenic (in poultry) 140
arteries 51
arteries, coronary 65
arteries, dilates 50
arteriosclerosis 65, 244
arthritis 54, 82, 128, 314
ascorbic acid 175
ASD 303
aspartame 21
aspartate 194
aspartic acid 194, 195–196
Association for International
    Cancer Research 81
asthenia 177
asthma 4, 14, 43, 45, 53, 82, 179,
    345
atheloid movements 97
atherosclerosis 12, 244
athletic performance 249
Atkins, Dr. Robert C. 153
ATP xvii, 4, 36, 49, 89, 98, 127, 128,
    193, 195, 231, 245, 250, 348
ATPase enzymes 299
ATP production 96
Attention Deficit Disorder
    (ADD) 181
Attention Deficit Hyperactivity
    Disorder (ADHD) 181
atypical antipsychotics 173
Australia 50

autism xiii, 20, 98, 121, 134, 229,
    308
Autism Research Institute 101
autism spectrum 99, 135
avian flu 217

B

bacteria 191
bacterial infection 4
Barnett, Dr. Lewis B. 58, 263
Beta-adrenergic agonists 28
beta-carotene 87
beta carotene 23
bio-chemical attack 223
bioavailability 203
biosynthesis, eicosanoid 96
bioterroism 223
Biotics Research Corporation 245
bipolar disorder 174, 176
birth control pills 27
birth defects 235
bladder 264
bladder infections 142
blast cells 92
Blaylock, Dr. Russell 13, 98
blood-cerebrospinal fluid barrier 191
Blood Brain Barrier 50, 172, 191
blood cells, granulozytic 90
blood pressure 37, 71, 166, 167, 204,
    244, 298
blood serum 2
blood sugar 3, 240, 254
blood vessels 264
body temperature regulation 273
bone mineral density 56, 272
bone mineralization 53
bones 57, 61, 264
borneol 113
Boston Globe, The 225
BOTOX® 151
botulinum toxin 152
botulism 152
bowel 301
brain 97, 163, 167, 172, 213–214,

263, 264, 265
brain, chemical imbalances in 349
bread, whole-wheat 27
breast cancer 260
breasts 264
breast tenderness 261
Brilla, Lorrie 252
bronchoconstriction 52
bronchospasm 53
Browne, Dr. S. E. 148
Bufferin 164, 165
Buttar, Dr. Rashid xiv

C

C. *diff* 15
cadmium 129, 139
caffeine 27
calcification 60
calcitonin 246
calcium 2, 30, 49, 58, 168, 195, 265,
      297, 298
calcium, balancing with
      magnesium 269
calcium, endothelial intracellular 244
calcium, unabsorbed 54
calcium channel blocker 244
calcium deficiency 51
calcium intake 273
calcium oxalate 179
calcium to magnesium ratio 348
Campbell, Jonathan 221
camphor 113
cancer xx, 16, 56, 81–82, 91, 93,
      229–231, 238, 260
cancer, breast 238
cancer, in children 82
cancer, metastatic prostate 55
carbohydrates 2, 137
carcinogenesis 90
carcinogenic compounds 83
cardiovascular disease 4, 45, 71, 231
Carque, Otto 57
carvone 113
catecholamines 242, 263

cation 1, 187, 188, 328
CDC 135, 136, 218, 224
cell division 91
cell membrane 242
Centers for Disease Control 134, 217
central nervous system 191
cerebrospinal fluid (CSF) 191, 213
cesium chloride 93
chelation 97, 103, 129
chelation, metal 108
chelator, magnesium not a 104
chelators 22
chelators, synthetic 130
Chelorex™ 115, 117–118, 352
chemotherapy 83, 92, 93
chicken pox 222
chilblains 43
children 193, 300
children, obese 136
China 66
China, preventive medicine 66
chiropractors 64, 198
chlorella 113–114
Chlorella Growth Factor 117
chlorella vulgaris 117
chloride 189, 211
chlorophyll 299–300
cholecystitis 43
cholesterol 50, 237, 240, 262, 316
choroid plexus 191, 192
chromium 255
chronic disease 82
Chronic Fatigue Syndrome 16, 45,
      74, 345
cilantro 112–113, 115
circadian 265
Clark, Dr. Hulda 63
Clark, Dr. Larry 87
claudication 148
clostridium botulinum 152
clostridium difficile 15
clotting, blood 16, 265
Clozaril 181
cocaine 28
CODEX xviii

Codex Alimentarius 26
Cohen, Lee 272
coldcure.com 178
colitis 43, 73, 299
colon cancer 73
Colorado, University of 54
conception 319
congestive heart failure 254
conjunctivitis 345
constipation 74, 142
contaminants, chemical 81
contamination, food 20
contraindications 198
convulsions 97
cooking, quick 68
copper 56, 100, 231
coriander 112
coronary heart disease 50
corticosteroids 28
cortisol 199
cortisone 262
Crohn's disease 34, 73
Crosby, Dr. Vincent 90
CSF 174
Curb, J. David 260
Cure for All Diseases, The 63
Cutler, Dr. Andrew Hall 112, 115
cytokines 95, 230
cytophilactic activity 346

D

dairy products 55
Davern, Dr. Tim 287
Davies, Dr. Stephen 11
Dead Sea 189
Dead Sea salts 312
Dean, Dr. Carolyn xvii, 3, 131, 167,
    293
death 319
death, iatrogenic xix, 11, 222
Defeat Autism Now 121
deficiency, magnesium 2, 9, 13, 19,
    103
deficiency, masked 28

Deger, Orhan 235
degranulation, cell 52
dehydration 133, 300
dehydration, intracellular 133
Dehydroepiandrosterone (DHEA)
    238, 262
Delbet, Dr. Pierre 41, 42
delirium 177
dementia 177
dental research 215
deodorant 205
depression 4, 5, 74, 100, 163, 171,
    173, 177–178, 300
detoxification 96, 103, 128, 129
detoxification, heavy metal 97
detoxification, phase one 104
Dextrorphan 166
DHEA xiv, 199, 237, 238, 239–242,
    260, 267, 271
DHEA, synthetic 238
DHEA-Sulfate 240, 262
DHEA and Aging 242
DHLA 111
Diabeta 302
Diabetes 4, 5, 82, 131, 181, 229,
    264, 305, 316
diabetes xx, 238, 279
diabetes, Type I 133, 153
diabetes, Type II 12, 70, 132–133,
    153
diabetes mellitus 132, 140, 147
diabetic foot 145
diabetic ketoacidosis 132
diabetic neuropathy 137
diarrhea 29, 142, 165, 180, 201, 350
diet, American 30
Dietary Reference Intake 298
digestion 350
digestive system 64
dihydrolipoic acid 111
dioxins 196
diuresis 133, 273
diuretics 27, 28
dizziness 142
DMPS 109, 118, 130

DMSA 109, 118, 130
DNA 91, 127
dodecanal 113
dopamine 242, 263, 269
dosage 202
double helix 3
Downing, Dr. Damien 11
Doxycycline 302
drug companies 183
drugs, psychiatric 173
Durlach, Dr. Jean 209
dysrhysthmia, cardiac 180

E

Eby, George 178
eclampsia 71, 168, 301
eczema 43
EDTA 109, 118
eicosanoids 95
elasticity, vaginal 267
electrical balance 2
electroencephalogram 5
electroencephalogram (EEG) 265
electrolytes 255
electrolytes 263
elemol 113
elephant tusks 57
Eli Lilly 175
Elin, Dr. Ronald 68, 72
emphysema, pulmonary 345
endocrine 259
endocrine system 3
endometrial cancer 261
endometrium 265
endothelial dysfunction 230
endothelium 230
endrocrine function 265
energy 4
Environmental Protection Agency
    (EPA) 140
enzyme reactions 190
enzymes 3, 5, 16, 127, 243
enzymes, magnesium dependent 3
epigenin 113

epilepsy 100
Epsom salts 209, 252–253, 312–313
equimolar ratio 117
Ermakova, Dr Irina 141
erysipelas 345
erythromelalgia 149
esophageal cancer 86
estradiol 265
estrogen 238, 242, 259, 261–266
estrogen, low 245
estrogen-progesterone imbalance
    245
estrogen therapy 259
evaporation, solar 189
Excedrin® 287
excitotoxicity 194
exercise 63
extracellular fluid 191
Extreme Sports Medicine 252
eye washes 190

F

facial hair 238
fascism, medical 92
fatigue 4, 63
fatigue, chronic 128
Fauci, Dr. Anthony S. 222
FDA 30, 92, 134–135, 141, 152, 219,
    224
fecundation 236
fertility 235
fertilizers 25
fetal tissue 235
fetus 235
fibromas 5
fibromyalgia 45, 73, 128, 178, 347,
    350–351
Ficks Law 202
Fink, Dr. Matthew E. 139
Finland 65–66
Finland, magnesium intake in 65
Fisher, Dr. Leslie 121
flavonoids 113
flax seed 269

fluid retention 268
FluMist™ 223
fluoridation, water 83
fluoride 4, 28, 58, 139, 151
Flu Prevention 222
folic acid 175
Food and Drug Administration
    (FDA) 21, 76
food industry 139
foods, genetically modified 141
foods, processed 72, 99
foot bath 200–201, 254
foot soaking 37
foot ulcers 145, 151–152
Foster, Dr. Harold 85
free radicals 192, 230
French, Claudia xiv, 261, 277
fructose 350
fructose level 143
FSH 263
Fuchs, Dr. Nan Kathryn 25
fungal infections 5

G

GABA 263
Galland, Dr. Leo 175
gallbladder cancer 73
gangrene 145, 148
gastric cancer 90
gastroenteritis 299, 345
genitals 244, 264
genital soreness 267
gentamicin 113
Geodon 181
Georgiou, Dr. George 112, 115, 120
geraniol 113
Gilead Sciences 225
Gilliland, Frank D. 68
gingivitis 190, 315
Gittleman, Ann Louise 269
GLA 152
glaucoma 4
glial tissue 110
glipizide 302

glucose 132
glucose, high blood 143
glutamate 194
glutamate removal 194
glutamic acid 175
glutathione 44, 95, 110, 117
glutathione, intracellular 96
glutathione production 15, 129
glutathione synthetase 98–99
Glyburide 302
Goldsmith and Baumberger 267
Gonadotropin Releasing Hormone
    265
Goodman, Dr. Sandra 83
Gordon, Dr. Garry xiv, 49, 96,
    108–109
grains, refined 69
granulocytes 92
Great Salt Lake 189
Greenberg, Dr. Alan 110, 120, 184
gums 202, 215

H

hair 264
hair health 190
Haley, Dr. Boyd 121, 128, 196
Hall, Dr. Andrew Cutler 121
hallucinations 177
hamstring, shortened 254
Harvard Medical School 12, 272
Harvard University 73
hay fever 43
headaches 14, 287
hearing loss 5
heart, the 64, 240, 264
heart attack 4, 45, 50, 71, 135, 151,
    164, 238, 254
heart disease 16, 82, 229, 238, 264,
    279
heart failure, congestive 10, 254
HeartHealth 63
heart rate 301
heart rhythm 3
Heavy Metal Detox 116

heavy metals 97–98, 196, 272
Hegerty, Nannette H. 171
hepatitis B xvi
Herman, Donald 140
herpes zoster 345
high blood pressure 16, 71, 279, 316
Hillier, Dr. Teresa A. 135
hippocampus 214
HMD™ 118–120
Hoffer, Dr. Abram 176
Hoffman, Dr. Ronald 34
Homeostasis 255, 299
homicides 172
hormone 63
hormone, master (DHEA) 242
Hormone Replacement Therapy
    260–261
hormones 260, 263, 269
hormones, sex 234
hormones, transmission of 242
hot Flashes 259, 263, 269, 273
Howenstine, Dr. James 87, 164
HRT 260–261
hydration 68
hydrochloric acid 32, 190, 211
hydrocortisone 262
hydrogen peroxide 91
hydroxylation 22
hyperarousal 177
hypercalcemia 56
hyperglycemia 133, 143, 148
hypermagnesemia 133, 200, 299,
    300
hypertension 133, 151
hyperthermia 273
hyperventilation 177
hypocalcemia 59, 60, 147, 301
hypocalciuric hypercalcemia,
    Familial 300
hypoglycemia 143, 193, 264
hypokalemia 60, 147
hypomagnesemia (magnesium
    deficiency) 59, 132, 148, 235,
    264
hypomania 178

hyponatremia 133
hypothalamus 263, 265
hypothyroidism 299–300
hysterectomy 261

I

iatrogenic death 75
imbalance, hormonal 260
IMC Research Laboratory 197
immune function 16, 77
immune system 17, 91, 128, 191,
    200, 231
impotence 5, 43, 142
incontinence 264
infarction 88
infectious diseases 44
infertility 63
inflammation 229, 305
influenza 200, 224, 345
influenza epidemic 219
influenza pandemic (1918) 217
insomnia 4, 74, 100, 174, 308
insulin 91, 131, 147, 263
insulin, synthetic 144
insulin resistance 70, 136
insulin secretion 269
intake 19, 53, 201, 225, 255
intake, excessive calcium 55
intake, magnesium 12
intercourse, sexual 267
International Medical Veritas
    Association xiv, 222
interstitial fluid 191
intestine 56
Intracellular Free Magnesium 293
intracellular ion 289
Intracellular Magnesium Test 37
intracranial 192
intramuscular injection 35, 71
intravenous 30, 39, 71, 301
ionic bonds 188
ionic compounds 187
iron 56, 69
Ironman 285

irritability 173, 177, 268
Irritable Bowel Syndrome 350
ischemic heart disease 65

J

Japan 50
Johnson, Steven 4
Johnson, Sylvester M. 172
Journal Environmental Health
    Perspectives 140

K

kaempferol 113
Kaiser Permanente 243
Kaye, Dr. P. 52
Kentucky, University of 196
ketoacidosis 132
kidney 299
kidney disease 82, 133
kidneys 4, 51, 53, 64, 128
kidney stones 16, 179, 287
Klenner, Dr. Frederick R. 221, 222,
    225
Krebs, Dr. Nancy F. 54
Krebs Cycle 253
Kubena, Dr. Karen 53

L

L-Glutamine 117
Laumann, Dr. Edward 236
laxatives 301
lead 129, 139, 194
leaky gut syndrome 100
Lee, Dr. William 287
Leone, Dr. Nathalie 231
lethargy 173
leucocytic activity 189
leukemia 41
leukocyte 230
Levy, Dr. Thomas E. 50
LH 263
libido 234
life, last stage of 237

Life Extension Foundation 30–31,
    241
limonene 113
linalool 113
lithium 303
lithium therapy 300
Liu, Dr. Simin 12
liver 38
longevity 256
Lou Gehrig's Disease 74
lubrication, vaginal 244, 267
lung cancer 87
lungs 77, 192
lupus 128, 238
lymph fluid 128
lymphocytes 92
lymph system 128

M

macronutrients 2
macrophage 230
maginex 149
magnesium 1, 298
magnesium (in pregnancy) 234
Magnesium, The Miracle of xvii
Magnesium, Total Red Cell 293
magnesium absorption 29
magnesium aspartate 90
magnesium carbonate 253
magnesium chloride xix, xx, 9, 10,
    187, 189, 198, 222, 225, 238,
    243, 252–254, 313
magnesium chloride, transdermal
    271
magnesium chloride hexahydrate 31
magnesium citrate 253, 350
magnesium deficiency 28, 64, 96,
    140, 174, 249, 289, 350
Magnesium Deficiency in the
    Pathogenesis of Disease 65
magnesium deficit 38
magnesium fumarate 253
magnesium gluconate 150
magnesium glycinate 253

Magnesium Loading Test 294
magnesium malate 253
magnesium oil 36, 197, 268, 299
magnesium oxide 350
Magnesium Research Institute 302
magnesium sulfate (Epsom salt) 10,
    13, 148, 209, 212, 253, 280,
    286
magnesium sulfate (IV
    after stroke) 167
magnesium supplementation 67
magnesium taurate 199, 203, 253
magnesium toxicity 299
Malatesta, Dr. Manuela 141
manganese 32, 56
Mansmann Jr., Dr. Herbert 15, 148
massage therapist 200, 206, 254,
    285, 318
master hormone 267
Mayhill, Dr. Sarah 14, 16 31, 34,
    255, 290
Mayo Clinic 219
Mazur, A. 230
McBean, Dr. Eleanor 219
measles xvi, 222
Medicaid 174
Medical Industrial Complex 256
medicine, preventive 66
MediSin 24
Medline 211, 261
Mefennamic Acid 303
melatonin 265
Membrane Permeability 202
memory 214
Memory and Cognitive Function 191
Menopause 259, 264, 266, 267, 269,
    270, 272
menstrual cycle 243
menstruation 270
mental retardation 173
mercury 20, 103, 112, 117, 128, 139,
    192, 193, 194, 197, 225, 255
Mercury, The Rising Tide of xiv
mercury amalgam 215
mercury poisoning xiv, 98

mercury selenium 197
Meston, Cindy 236
metabolism 2, 3, 128, 190
metallo-enzymes 172
metallothionein 109, 110
metastatic prostate cancer 55
methyl mercury 193
microalbuminuria 133
microfibrils 114
microglia 194
Micronutrients 2
Midol Teen Formula 287
Migraine, Menstrual 271
migraines 4, 14, 45, 70, 303, 349
milk 69
Milk alkali syndrome 300
milk lobby 61
minerals, declining levels of 23
minerals, trace 24
Minkoff, Dr. David I. 285
Minnesota, University of 88
minocycline 302
MIT 268
mitochondria 52, 245
mitochondrial function 95
mitral valve prolapse 74, 349
modern medicine xix
monocarboxylic acids 193
mononucleotides 3
monosodium glutamate (MSG) 195
mood disorders 100, 174, 272
Morris, Dr. J. Anthony 219
mortality, infant 235
mortality rates 82
Mother Earth News 23
mouth cancer 73
mouthwash 190, 202
MSM 115, 117
mucosa 32
mucous membranes 202, 264
Multiple Sclerosis 4, 74, 100
mumps 222
muscle 252
muscle relaxant 52
muscle relaxation 16

muscle spasms 60
muscle spasms, night time 178
Myalgic Encephalomyelitis 314
Myers' cocktail 45
myocardial infarction, acute 13
myocardial infarction, infantile 65
myocardial infarctions 10, 39, 65, 142, 148, 166

N

N-methyl-D-aspartate 166, 195
NAC 117–118
Nadler, Dr. Jerry L. 131
National Foundation for Cancer Research 85
National Institute for Occupational Safety and Health 83
National Institute of Allergy and Infectious Diseases 222
National Institutes of Health 176
Nature's tranquillizer 270
naturopath xv
nausea 261
NCAA 252
NDF 119
NDF Plus 119
nephropathy 132
nerve function 77
nervous system 3
neuritis, optic 345
neurodegenerative diseases 194
neurons 195
Neurontin 302
neuropathy 90, 143, 148, 151
Neuropathy, Diabetic 141
neurotransmitters 172, 195, 242, 263
Neveu, Dr. A. 39, 40, 191, 345
New England Journal of Medicine 238
New York Academy of Sciences 242
New York Times, The 217
New Zealand 50
niacinamide 175
nicotine 28

Nieper, Dr. Hans A. 90
NIH 178
nitric oxide 99, 143, 244
nitrogen mustards 83
NK cells 92
NMDA-receptor 100, 168, 214
NMR 293
North Carolina, University of 279
Northrup, Christianne 269
Northwestern University School of Medicine 12
Nottingham City Hospital 90
N P K fertilizer 24
NSAID 303
NutraSweet® 195
nutrition 3, 5, 17, 21, 25-26, 51, 53, 59-60, 69-70, 76, 83-84, 99, 117, 137, 139, 152, 174, 175, 179, 183, 190, 222, 231, 249, 254, 256, 280, 286, 305, 326
NyQuil Cold and Flu 287

O

O'Toole, Dr. Tara 218
obesity 134, 183, 242
Obsessive Compulsion Disorder 176
Office of Women's Health 268
Omura, Dr. Yoshiaki 112
oral 30
Oral Magnesium Supplementation 253
organelles 2
orgasm 243
Ornstein, Robert 243
osmotic action (magnesium loss thru) 133
osteomyelitis 345
osteoporosis 4, 53, 56–58, 61, 264, 272, 279
Osteoporosis Society, National 57
osteosarcoma 83
ovarian cancer 73
ovary 265
ovule maturation 236

oxalic acid 32
oxyradical scavengers 96

P

pain, reduction of 286
pain relief 190
paired minerals 51
pancreas 139, 140
pandemic 217
panic attack 4
paralysis 163
paranoia 173, 179
parathyroid disease 56
parathyroid hormone 59, 265
parenteral administration 191
Parkinson's disease 43, 100, 238
Pasteur Institute 231
patellar reflex 300
Patrick, Dr. Lyn 110
Pauling, Dr. Linus 45
Pavia, University of 141
PCBs 196
PDR 302
pediatricians 54
Pedula, Kathryn L. 135
pelvic muscles 264
penumbra 167
Percocet 287
perimenopause 260
perimenopause 263
Periodontal Disease 213
periodontitis 214
peroxynitrite 99
peroxynitrite damage 4
pesticide 99
Pesticide Action Network 21
pesticides (in foods) 141
phagocytosis 189
Pharmaceutical Industry 256
PHN 286
phosphatase 60
phosphate of magnesium 57
phosphates 28, 348, 350
phosphorus 58

Physicians for Social
    Responsibility 82
phytochemicals 113
pineal gland 265
Pinto, Dr. Al 168
pituitary function 245
pituitary gland 263
placebo 252
placental tissue 235
plasticity 214
platelet aggregation 164
PMS 128, 268, 303
PMS, borderline magnesium levels
    269
pneumonia, viral 222
poisoning, alimentary and
    professional 345
poisoning, chemical 82
Poland, Dr. Gregory A. 219
polio 222
postmenopause 241
potassium 25, 51, 96, 97, 133, 195,
    298
poultry industry (use of arsenic) 140
pre-diabetic 134, 147
pre-eclampsia 235
preeclampsia 71
pregnancy 71, 234, 235, 263
pregnenolone 262
Premenstrual Syndrome 259, 268
prenatal 71
preventive medicine 66
prickly sensations 179
progesterone 260, 262, 263
Progestin 261
prohormone 242
Prostaglandin E 153
prostate cancer 85, 86, 87
prostatic hypertrophy 43
protein 129
protein synthesis 3
proteinuria 132–133
psoriasis 43, 128
psychotropic drugs 182
PTH 59

puberty 237
Pusztai, Dr. Alpad 141
pyridoxine 175

Q

quadriceps 254
quercitin 113
Quesnell, William R. 60

R

radiation 83, 93
rage thing, the 171
Rapson, Dr. Linda 278–279
rash 308
Rath, Dr. Mathais 22, 92
ratio, magnesium to calcium 351
Raynauds phenomenon 349
RDA 55, 77, 180, 231, 250, 251
re-mineralization 45
Recommended Dietary Allowance 5
red blood cell production 240
refining grains 68
Reid, Daniel xiii
Relenza 219
renal disease, end-stage 133
renal diseases 65
renal failure 29, 300
renal insufficiency 299
Resnick, Dr. L. M. 148
restless leg 287
retinopathy 132
rhamnetin 113
rhinitis 45
Rimland, Dr. Bernard 101
Risperdal 32, 181
Ritalin 181
RNA 127
Robinson, Jo 23
Roche Laboratories 225
Rodale, J. I. 40, 268
Rodriguez, Dr. Carmen 55
Rogers, Dr. Sherry 120
Rude, Dr. Robert 68, 72
Rumsfeld, Donald 225

S

salbutamol 53
salmonella 113
schizophrenia 174, 176
Schutt, Jeff 254
sciatica 287
Science Formulas 115
scurvy 221
seawater 42
Seelig, Dr. Mildred 25, 34, 65, 245,
    266, 279, 280, 320, 323, 331-
    332, 336
seizures 100
selective serotonin reuptake
    inhibitors 173
Selenium 255
selenium 68, 69, 81, 85, 117, 140,
    197
selenium deficiencies 86
Semen transport 244
Senegal 86
senile tremors 43
Seroquel 181
serotonin 242, 263
serotonin deficiency 172
serotonin synthesis 270
serum calcium concentration 60
Serum Ionized Magnesium 293
Serum magnesium 148
sex 233, 243, 245
Sexual Dysfunction, Female 236
sexual energy 190
sexuality 233
Shands Cancer Center, Univ. of
    Florida 56
Shealy, Dr. Norman xiv, 35, 37, 199,
    267
S.I.D.S. 20
Silbergeld, Dr. Ellen 140
sinuses 192
sinusitis 45
sinus problem 128
skin 264
Sobel, David 243

sodas 29
sodium 29, 195, 298
sodium, retention of 60
sodium chloride 133
soil 25, 27, 99, 251
sorbitol 143
soya 141
soy beans 68
Spanish Influenza epidemic of 1918
    221
spasming, coronary arteries 65
spasms, muscle 201
sperm viability 236
spirulina 114, 298
sporopollenin 114
sport injuries 201
sports medicine 249, 255
sports nutrition 190
Standard American Diet 294
Stanford University 85
statin drugs 151, 167
Steckel, Mark 253
steroids 251, 262
Stoll, Dr. Walt 39
stomach cancer 73
stress 175, 237
stress, magnesium loss due to 350
stress, systemic 229
stroke xx, 13, 151, 163, 164, 167,
    193, 260
stroke prevention 164
Sublingual Magnesium Assay 294
sudden infant deaths 65
sudden infant death syndrome 20
sugar 25, 27
suicide 5, 172
suicides, teen 182
sulfate 209, 211
superoxide 99
superoxide dismutase 91
suprachiasmatic nucleus
    (hypothalamus) 265
sweat 255
sweating 29
symptoms, magnesium deficiency
    176
Synapses 213, 214
systemic stress 229
Szasz, Dr. Thomas 183

T

T-cell 112
T-cells 92, 230
T-lymphocytes 91
Tamiflu 217, 218, 220, 222, 224
taurine 117
TD-DMPS xv
teeth 57, 190, 202
testimonials 305
testing 289
testosterone 234, 238, 239, 240, 242,
    251–252, 262, 263
tetanus 222
tetany 60, 301
tetracycline hydrochloride 302
Texas, University of 287
Texas A & M University 53
thaler (early Roman currency) 92
Theophylline 28
Theraflu 287
therapeutic index 36
thiamine 3, 175
thiamine deactivation 5
Thiazide 28
thimerosal 65, 98
thinking, destructive 70
Thomas, Dr. David 69
Thor, James 252
thromboembolic events 261
thrombophlebitis 148
thrombosis, acute coronary 10
thromboxane 244
thyrocalcitonin 61
thyroid 242, 245, 263
tingling, full body 177
tinnitus, 124
tissue saturation 38
tocolytic magnesium sulfate 210
tonsillitis 345

tooth cavities 5
tooth decay 53
tooth enamel 190
Tourette's syndrome 176
toxicity. heavy metal 129
toxification 134
toxins 313
Track and field 250
transdermal 17, 30, 33, 36, 39, 129,
        151, 168, 190, 198, 249, 253,
        271
transdermal magnesium (natural
        DHEA) 239
transdermal use 267
trembling 174
tremor 60, 302
trials, double-blind 269
triathalon 285
triglycerides 60
triphosphate 127
tryptophane 175
twitching 174
Tylenol® 287

U

ulcer, peptic 73
ulceration 145
ulcers, foot 148
Université P. et M. Curie 209
University of Chicago 236
University of Hawaii 260
University of Helsinki 297
University of Southern California 68
University of Texas at Austin 236
uptake, mineral 32
Urbino, University of 141
urethra 264
urinary organs 264
urinary tract infections 264
urine 25, 129, 133, 255
urticaria 43
US population, magnesium
        deficient 147
uterine cancer 260

uterus 261
uterus 264

V

vaccination 65, 219
Vaccine Injury Compensation
        Program 220
vaccines 20, 98, 220, 225
vaccines, childhood 65
vagina 190, 264
vaginal application 243
vaginal bleeding 261
vaginal dryness 264
vaginal mucus 265
vascular degeneration 60
vascular resistance 244
vasodilation 244
vasospasm, cerebral 168
vegetables, green 29
ventilation, artificial 300
Vergini, Dr. Raul 42, 204–205
vertigo 97
Viagra 243
Vicodin 287
Violence 171
viral diseases 191
Vitamin A 65
Vitamin A and Zinc 91
Vitamin B 97, 100, 129, 175
Vitamin C 16, 22, 65, 68, 97, 110,
        117, 129, 176, 201, 221–222,
        224–225, 238
Vitamin D 33, 60–61
Vitamin E 68, 72, 97, 129
VLDL 60
vulva 264

W

"War on Diabetes" 76
Walker, Dr. Sydney 183
Warfarin 166
warts 43
water, fluoridated 83
water retention 260

Waters of Life 298
weakness 142
Weil, Dr. Andrew 84
Werbach, Melvyn R. 269
Western Washington University 252
Whitaker, Dr. Julian 182
Whitaker, Dr. Scott 24
Whitaker, Robert 173
white blood cells 40
White House 223
Womens Health Initiative 260
Wong, Dr. Cathy 286
World Health Organization xviii, 81
wrinkles 190

Y

Yazbak, Dr. F. Edward 219
Yu, Dr. Shu-Yu 86

Z

zinc 56, 68–69, 81, 91, 117, 172,
        176, 192, 231, 251–252, 255
zinc lozenges 222
Zoloft 178
Zyprexa 32, 181

Additional Titles from Phaelos Books
Available online at phaelos.com

I Am My Body, NOT!
ISBN 0-9700209-1-0, 60 ppg., hardcover, $19.95
— Written by Adam Abraham, illustrated by Marie Litster. The comprehensive introduction to the workings of the human body is really only a beneficial by-product of this ground breaking book. It's real value lies in the consistent, conscious, and loving message that it sends about the importance of love, permanence of life, and of who we are.

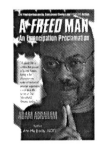

A Freed Man: An Emancipation Proclamation
ISBN 0-9700209-0-2, 304 ppg. softcover, $17.95 — Written by Adam Abraham. A collection of twenty-one essays that present a blueprint for positive individual and social change, beginning within, through what the author calls Wonderful Dreams.

Contact Information

Phaelos Books & Mediawerks
860 N. McQueen Rd #1171
Chandler, AZ 85225-8104

480.275.4925
509.479.8415 (fax)

www.phaelos.com
info@phaelos.com

What is PHAELOS?

Our name PHAELOS (*fy-los*) is derived from the Greek, *philos* that means fondness, and love. Phaelos is a natural quality and expression of humanity that we believe is not necessarily in short supply, but under-expressed—if not repressed— in our society.

Phaelos Books was inspired by a vision of, desire for, and commitment to creating a better world through conscious, loving, and fearless choice—one mind and *heart* at a time—through self-reliance and positive cooperation among people.

Phaelos is *a passion for humanity*™.

CPSIA information can be obtained at www.ICGtesting.com
Printed in the USA
LVOW110951210513

334582LV00002B/18/A